The Quest for
Life in Amber

The Quest for
Life in Amber

George and Roberta Poinar

§ Helix Books

♦ Addison-Wesley Publishing Company

*Reading, Massachusetts   Menlo Park, California   New York
Don Mills, Ontario   Wokingham, England   Amsterdam   Bonn
Sydney   Singapore   Tokyo   Madrid   San Juan
Paris   Seoul   Milan   Mexico City   Taipei*

Many of the designations used by manufacturers and sellers to distinguish their products are claimed as trademarks. Where those designations appear in this book and Addison-Wesley was aware of a trademark claim, the designations have been printed in initial capital letters.

Library of Congress Cataloging-in-Publication Data

Poinar, George O.
    The quest for life in amber / George and Roberta Poinar.
      p. cm.
    Includes bibliographical references and index.
    ISBN 0-201-62660-8
    ISBN 0-201-48928-7 (pbk.)
    1. Fossils.   2. DNA, Fossil.   3. Amber.   I. Poinar, Roberta.
  II. Title.
  QE742.P66   1994
  560—dc20   94-3043
                                          CIP

All acknowledgments for photographs can be found on page 218

Cover design by Lynne Reed
Text design by Janis Owens
Set in 9.5-point Stone Serif by Compset

1 2 3 4 5 6 7 8 9 10-MA-9998979695
First paperback printing, September 1995

"The scientist does not study nature because it is useful. He studies it because he delights in it and he delights in it because it is beautiful."

Jules Henri Poincaré
1854–1912

Contents

Preface

Millions of years ago, trees from now vanished forests produced deposits of resin that transformed into what we know today as one of nature's most beautiful gems, amber. The forces that caused the sticky resin to slowly harden remain a mystery. While still viscous, the resin acted like flypaper and tenaciously held insects, plant parts, and even small vertebrates that touched it. Eventually the partially hardened yellow clumps, complete with their assemblage of enclosed organisms, fell from the trees to the earth. There, first leaves and debris, and later soil and rock, covered them, and eventually they were buried in layers of sandstone, limestone, or even coal. All this time, the original resin was developing the hardness, density, and melting point characteristic of amber. This durable organic gemstone outlasted the parent trees, the forests they formed, and the ecosystem of which they were a part. In some places the rock layers containing fossilized resin settled beneath the sea, and, over eons, the currents slowly loosened the amber from its grave. On the sea bottom, the amber was exposed to marine creatures that would even grow on the surface of larger pieces. Eventually the loosened amber was washed up on beaches, where humans eagerly competed with each other to collect this gold from the sea.

In other regions, the amber-bearing rock layers were shoved up into mountain ranges, where release of the gem depended on the forces of erosion. Where amber is locked up in layers of the earth, humans have dug mines to locate the "veins" and follow them as far as they can.

Whether because of its color, its feel, the mystery surrounding its formation, or the fascinating insect or plant remains it contains, amber is, and has been for centuries, valued by people. Any substance that can weather the forces of time must have magical powers. Just behold the sunlight as it reverberates off the internal fracture planes of raw amber and you will understand why many people consider it mystical.

Because amber is highly desired, it is valuable and has been a commodity of trade and sale for thousands of years. But the collecting and selling of amber has its dangers, and many have paid for this activity with their lives.

My own interest in amber was sparked early in my life. My mother, Helen Louise, an avid reader, was especially fond of history and accounts of the early Greeks and Romans. Adjacent to her reading chair stood an end table filled with a constantly changing variety of books. Before I could master the words, she was reading bits and pieces from this or that book or explaining the main theme of the work to me.

Two books especially impressed me, the first dealing with the discovery of humans and animals that had been buried under the volcanic ash of Mount Vesuvius at Pompeii. I was stunned by how quickly these poor victims had been "frozen" in time as they conducted their daily activities. The second memorable book had a small figure printed in the upper corner of the title page—a weevil embedded in amber. I couldn't read the text, but my mother explained that such pieces really existed. Without knowing, and in her quiet, gentle way, my mother had presented me with two visions stamped forever into my memory and forming the inspiration for a portion of my life's work.

As a boy growing up in Ohio, I had no opportunity to discover any actual amber, so I had to be content with occasional visits to natural history museums. At that time all the amber I encountered came from the Baltic area, and I had no idea that I might one day visit northern Europe. The mystery and wonderment of amber never left me, although other activities took precedence during my high school and college years. The opportunity to work with amber finally presented itself after I had joined the faculty of the University of California at Berkeley. Once I initiated studies of insects, nematodes, and microorganisms, it wasn't long before I was thoroughly bitten by the amber bug and became hooked.

Roberta had already established her reputation as a leading electron microscopist at Berkeley, and when we began working together, we realized that now was the time to determine how well amber preserved the internal tissues of entombed insects. We were essentially on our own, since no one ever had attempted to section amber fossils for the electron microscope. We encountered many unexpected problems and setbacks, but at last, we were thrilled to look upon the oldest known animal cells with nuclei and organelles. This breakthrough led us onward to search for the molecular basis of life, DNA, in amber-entombed insects and plants. Again the process was slow and tedious, but with a fortunate turn of events, we finally achieved success. That amber could preserve DNA for over 100 million years was a stunning revelation. The discovery of nuclei and DNA in amber inclusions not only disproved earlier theories on the limited life span of this molecule but revealed exciting new horizons for the use of ancient DNA.

Where can we go from here? The discovery of ancient DNA in organisms in amber raises several questions, one of which is how long DNA can persist on the planet? Also can life now be re-defined to include extinct DNA sequences that may be capable of being incorporated into and expressed by living organisms? Can life forms from amber be cloned? Can we finally tap into the genetic diversity of plants and animals from ancient tropical rain forests? Can we use ancient DNA to follow the movements of life across and between continents? What can amber tell us about the enigmatic topic of the origins and extinctions of life?

This book is a chronological account of our adventures in the amber world, including travels to remote areas in search of amber, meetings with people along the way, and efforts to research and study organisms, cells, nuclei, and the oldest known DNA.

But it is even more than that. This story shows how an idea can be presented to a child, undergo a period of dormancy and finally germinate into a burning interest. This interest, initially nourished by curiosity, later developed into a hobby and eventually into a scientific discipline leading to significant discoveries. All of this was made possible by unique circumstances, beginning with a love of natural history and amber, Roberta's knowledge of electron microscopy and our son Hendrik's interest and enthusiasm in molecular biology. Although the first person singular refers to me, both Roberta and I share equal roles in the preparation of this work.

Acknowledgments

We would like to thank the many people that aided us throughout our quest for life in amber: miners and others obtained amber samples, the amber dealers brought us countless fossil specimens, colleagues helped us identify inclusions and analyze various samples, and molecular biologists provided supplies and time during the initial attempts to isolate DNA from amber inclusions. We would also like to thank our parents.

The Quest for
Life in Amber

1   Amber from the Sea

The clouds raced across the steel blue North Sea, and the dune plants bobbed up and down from the gusty winds. Broken slabs of concrete were a grim reminder of the activity that had occurred on the beach some seventeen years earlier when the Axis forces were fortifying themselves for an invasion that never came, at least not on this lonely western coastline of Denmark. But World War II artifacts were not what I was searching for as I strode along the beach in July 1962. It was that mysterious gem that the Danes called *rav*, better known to us as amber.

I had read in various reports that amber often washed up on the beaches along the Baltic and North Seas, usually as small particles but sometimes as large lumps, and that if you were lucky, you might find a piece or two. If you were exceedingly lucky, you could find a piece holding a wonderful prize—a 40-million-year-old insect, perfectly preserved. Although the odds were slight, I could feel my heart start to race just at the thought of finding a fossil in amber. The path from the guest house across the sliding, sandy dunes to the beach was desolate; not a soul was in sight. By finding a position where the sun reflected off the wet sand, I could scan the glistening surface and wait for that golden glint signaling that a piece of amber had washed ashore.

This tract of land, called Blaavands Huk, was, on that day, a forlorn, deserted piece of no-man's-land. The cold northwestern wind blew stiffly across the breakers, rain clouds suddenly blocked the sun, and a chill went down my back. The huge waves washed up on the white sandy shore, constantly eroding its surface. Wind-borne sand particles beat against me, and finally I took temporary refuge on the protected side of a row of small sand

The amber coast of western Denmark.

dunes, waiting for the wind to lessen. Branches of heather, dwarf willow, and crowberry plants bent and twisted in the wind almost to the point of breaking.

I found no amber that day, and a Danish family later told me that local amber collectors pass along the beach each morning and collect all the large pieces. To find such pieces, one had to rise early and reach the beach before them. So I set my alarm that night and began walking the beach at eight o'clock the following morning. To my dismay, footprints followed the water edge; someone had already searched the area. The next day I got up at the crack of dawn, but footprints and wheel tracks told me that someone with a small wheelbarrow had already been there. Now I was perplexed. How could collectors see amber in the dark? Even with a flashlight, they wouldn't be able to distinguish a piece of oxidized amber from the many similarly colored stones that littered the beach. On the following night I didn't go to sleep but waited in the dunes. Finally, out of the darkness loomed the figure of an amber collector. It was a calm night and the

An amber collector's cottage lies nestled among the dunes in western Denmark.

figure walked along, stooped down, picked up something, and strangely carried the object to his mouth. There was a faint clicking sound. What did that mean? The hand then either went into a sack or dropped the object and started searching for another. Was the person smelling the amber or tasting it, and what was that sporadic clicking?

After the figure left and daylight arrived, I asked a Dane walking his dog along the beach if he could tell me what was happening. He laughed and explained that collectors can distinguish a piece of amber from stone, broken glass, and other artifacts by striking it against their front teeth. The sensation received from amber is quite different from that of other objects, and thus the other senses are not needed.

The wind still had not reached its full force on that day in early July. So I returned to the beach, now having reconciled myself to not finding large lumps of amber, and I began to search for fragments that the collectors had left behind. Finally, there they were—small, irregular particles partially embedded in the wet sand. Many were rough and dark on the outside, but a few had fractured and revealed the beautiful yellow color that one associates with amber. I had now collected my first pieces of amber from the North Sea in Denmark. Was there anything in them? My hand shook so much from excitement that it was difficult to identify the small dark speck in one of the bigger pieces. My hand lens showed things sticking out of the object. Those things were segmented; they were legs that belonged to a minuscule fly! All that waiting and searching had been worthwhile. Our search for life in amber had begun.

Looking down at the small pieces in my palm, I thought how strange it was to find fossilized plant resin in the sea. The tree resin had taken about 40 million years to harden, turn to amber, be buried under meters of soil and rock, and become submerged in the sea, only to resurface finally as the wave action loosened the amber from its locked case fathoms beneath the waves. Curious how the amber was now returning to the land. We still don't know how amber is made—no one has been able to synthesize it, and the same questions that challenge us probably perplexed prehistoric peoples when they carved amulets and jewelry from pieces they collected some 15,000 years ago. We don't even know for certain what kind of tree produced Baltic amber. Over the years, various scientists have come up with no less than seven different plants that supposedly produced the resin that became Baltic amber. Although we have made some progress since Pliny the Elder, in the first century A.D., provided evidence that Baltic amber was produced from tree resin, we still have much to learn.

Looking down the beach that day, I recalled reading that this area served as the beginning of the western amber trade route from the North Sea to Rome. Imagine that thousands of years ago, the ancestors of those present-day Danish amber collectors who had consistently outwitted me were also collecting amber on the western coast of Jutland for eventual transport to southern lands. Exactly when the trade began is lost in prehistory. Members of the Norva culture (an indigenous nordic neolithic group of settled fishermen) were supposed to have been trading about 3500 to 2500 B.C., and certainly, as the amber beads in the Mycenae shaft graves prove, amber from the north had reached Greece by 1400 B.C.

What was so desirable about amber that drove ancient Greeks and Romans to trade wine, oil, bronze vessels, salt, tea, silk, and even gold for this commodity from the sea? Amber has always been used for adornment and for protection from evil forces. Imagine the wonderment of a Stone Age gatherer who first held up to the sun a freshly fractured piece of amber and saw the golden light broken into dazzling, sparkling arrays as it reflected off the internal fractures. This substance must surely have magical powers.

The most important commodity to the early peoples who started trading this material was not Roman coins, gaudy wares, or colored beads, but metal ores for arms. Little lead, tin, and copper existed for the making of bronze objects in the North, and these cultures were in danger of remaining in the Stone Age after southern Europe discovered bronze. In fact, the Bronze Age is said to have arrived in northern Europe as a result of the desire for amber. It was an extremely important item in the development of those cultures.

The amber trade with Rome began before the birth of Christ and reached its golden age during the period of the Roman emperors. In the imperial city of Rome, amber was most widely used for ornaments and jewelry. Items for adornment consisted of beads of various shapes for necklaces, buttons of many sizes, pendants (some with insects), and rings, but also popular were sculptures of animals and humans. Some enterprising members of the assembly went so far as to have amber seals made with their private insignia. Ladies of high society used amber pieces to remove lint from their dresses, taking advantage of the static electricity that builds up on a piece of amber that has been rubbed on cloth.

Archaeologists have studied, discussed, and argued over the location of the amber trade routes for many years. All agree that such routes existed, and that one important route began on the western coast of Denmark.

The preferred path of the western route followed the river Elbe to the Danube, then over to Venice and south to Rome. The amber was placed in large sacks made of plant fibers and transported by boat or oxen cart. Imagine how nervous the drivers must have been, carrying such a valuable cargo. Not only did the merchants have to pay a duty on their treasure every time they entered a different province, they had to be on guard constantly for thieves. Many probably were robbed and killed before delivering their precious cargo. In fact, some must have foreseen disaster and quickly tried to conceal the amber. The only place to hide it was in the ground—if they could dig a hole and quickly cover it again before being detected. That some were successful in burying amber was shown by the discovery, in the nineteenth century, of caches of amber jewelry, badly oxidized with age, at various points along the ancient trade routes. Some of the amber articles were completely finished; others still required carving or shaping, and obviously were destined for some craftsman in Rome who had been commissioned to prepare a special work of art, perhaps a casket or sculpture. That some amber was never claimed hints at a fatal end for many amber merchants. But these unclaimed amber caches today provide solid evidence of business operations thousands of years ago.

At some point the Romans must have grown impatient with the banditry that interrupted the flow of amber, since they decided to go to the source themselves. In A.D. 5–6, they boarded small boats and traveled up the Elbe all the way to Jutland. This occurred even before they had discovered the Baltic Sea. They apparently did obtain some amber, but after experiencing great difficulty navigating up the river and dealing with the natives, they decided after all to wait for supplies to arrive from the North.

After returning to California in 1962, I crudely polished the pieces of amber I had collected on the Danish shore. The fly was the only fossil in the handful, but it was amazingly well preserved, and as I stared at it through my microscope I recalled a book with a drawing of a weevil embedded in amber that my mother had shown me many years before. If my interest in amber had been cursory, it now became intense. I began to collect and read as much as I could about amber and the fossils it contained. And I grew determined to plan as many trips as possible to amber-producing areas.

2   The Russian Connection

In 1969 I was awarded a NATO exchange grant allowing me to visit the then Soviet Union and discuss common research projects with interested Soviet scientists. The cold war was deeply entrenched, and the Vietnam conflict was straining relations between the United States and the Soviet Union even further; but this was an opportunity to meet Soviet scientists, see their research facilities firsthand, and possibly arrange some joint collaborative projects. Of course, I did have an ulterior motive. Before World War II, the largest collection of amber-bearing fossils in the world—some 120,000 pieces—had been maintained at the University Geological Institute Museum in Königsberg, at that time in northern Germany. After the war, that region fell under Soviet rule and Königsberg was renamed Kaliningrad in the district now known as Kalinin in Russia. But what had happened in the meantime to all that amber? Had it been destroyed, shipped out to other repositories before the battle of Königsberg in 1945, or confiscated and brought back to the Soviet Union? Maybe this question could be answered during my visit.

On a mild and pleasant April 1, 1969, in Amsterdam, I boarded an Aeroflot flight to Moscow for my first venture behind the Iron Curtain. A solid layer of clouds shrouded Moscow, and a blanket of snow covered the city. It was three degrees above zero when we landed, and the passport check, health inspection, and customs clearance agencies were all located outside the airport.

All passengers from the West were placed in the hotel Warshava, a stately old building reserved for foreigners. Guards at the door kept the local citizens from entering. After being whisked through a small group of

Russians clustering around the entrance, I was given a key and taken up to the fourth floor on a rickety old elevator. We were told in English that we shouldn't open the door of the elevator until it stopped completely; otherwise something horrible would happen. We could only imagine what that might be.

The room was luxurious by Russian standards—large, with a private bathroom and a window overlooking the streets of Moscow. I remember staring out at the myriad dark figures trudging patiently through the snow and slush. Most of the traffic consisted of taxis, delivery trucks, and emergency vehicles.

I spent much of the following day in my hotel room reviewing Russian vocabulary and grammar. I had taken six months of Russian at the University of California before departing, but on arrival I quickly learned that Russians didn't pronounce their sentences as clearly and precisely as the native speakers on the language tapes. A dictionary became my constant companion, and by the end of my visit the pages of that book were worn.

My official stay was to begin in Leningrad (now Saint Petersburg), but before leaving Moscow I was shown the Kremlin and Lenin's tomb. Both were interesting, but Lenin in his tomb, perfectly embalmed and honored by thousands of Russians daily, was truly extraordinary. Always a long line of people (something one grows accustomed to in Russia) waited to see this national hero. Before I entered the mausoleum, a guard took my camera and examined the contents of my pockets before waving me along. The line moved quickly. Inside the mausoleum more guards constantly inspected us, as soldier ants would inspect newcomers to their colony. Finally we descended into a low, stuffy hall. Still another guard removed my hat and ushered me into a dark room where overhead floodlights beamed down on Lenin's body. He was dressed in a black suit, with only his pale face and arms showing above a blanket.

We passed quickly around his body and entered another small room. Guards throughout kept saying "shhhh" to remind us that no noise would be tolerated. A workman in front of me started to put on his hat, but a guard whispered something in his ear and the workman quickly removed it again. Once back on the street, I realized that I had seen one of the greatest heros of the Russian Revolution.

The night train from Moscow to Leningrad left at midnight and took eight hours. The ride was bumpy, but the sleepers were clean and glasses of hot tea were a welcome surprise. Many consider Leningrad the most interesting of the larger Russian cities. Situated along the banks of the beautiful Neva River, it has numerous museums and institutes and probably a concentration of more scientists than any other Russian city. And because of its location and history, I felt a freedom and an openness that were lacking in Moscow. I never suspected I was being watched or followed in Leningrad, as I had felt in Moscow.

The next morning I left the hotel and walked down Nevsky Prospekt toward the Zoological Institute, which would be my workplace for the next three months. Most people had no car and the buses were literally packed, so walking seemed the most practical way to travel around the city. When I did take a bus, I often lost a button off my coat while forcing myself through the crowd to reach the exit door at my stop. But even the sidewalks were crowded with people, especially during the rush hours. All I could see were Russian coats and hats on all sides on my way toward the impressive brick Soviet Academy of Science building that housed the Zoological Institute.

On my first day, the guard at the door wouldn't let me enter and I had to wait until someone could verify my status as a temporary worker. I waited at the entrance of a large room with old wood panels lining the walls and ceilings. Finally someone escorted me to an upstairs office filled with scientists. The Russians were very pleased to see me and made every effort to make me feel at ease. Of course, they had dozens of questions about American and Western policies that needed to be answered as diplomatically as possible. One memorable question was whether it was true that all drinking water in the United States was polluted and our supply was imported from Canada.

Several opportunities arose during the next few months for me to learn more about Russian amber. Luckily, a large collection of fossiliferous amber was stored at the Zoological Institute. In fact, part of the institute housed a public museum with several old wood and glass cases filled with assorted pieces of Baltic amber. These cases stood next to a large glass container that held a stuffed baby mammoth salvaged from an ice field in Siberia. Several weeks after my arrival, an entomologist who had heard of my interest in amber took me aside and ushered me into a long narrow room. There two technicians were sorting through drawers and boxes filled with all sizes of amber pieces, each containing one or more insects. My eyes widened as I looked over what must have been several thousand pieces of beautiful Baltic amber. Smiling, my guide reached into a pile, pulled out two pieces, and handed them to me as souvenirs. Later I discovered that each contained a small fly.

But my friend couldn't tell me if any of these pieces came from the famous Königsberg collection. Russia had its own amber sources along the Baltic coast, and the government had recently reactivated the old mines in Kalinin that were near the coast. They had been flooded during the war, but were now drained and producing several hundred pounds of amber daily. The amber being sorted in the institute probably came directly from these mines.

Yantar, Russian for amber, was a common household word in Leningrad. Every family I visited had some memento or piece of jewelry made of amber—maybe a beautifully carved brooch, or a string of beads, or even a rosary that had been handed down from generation to generation as a family treasure. Most local museums, and especially The Hermitage, displayed many artistic objects made entirely, or in part, of amber. While attending the service at the Saint Nicholas Cathedral on Easter Sunday (a risky act for the Russian people at that time), old women wearing babushkas of all colors, their faces and hands weathered from laboring in the fields, filed slowly past me to briefly kiss an amber crucifix. This scene—with the gray-bearded priest waving a silver chained container of burning incense, the choir singing with such conviction and beauty that they might have been angels, and the pure devotion stamped on the faces of those women—moved me deeply.

People told me that during the siege of Leningrad (September 1941–January 1944), when thousands died from starvation, the family amber pieces were the last items to be pawned for bread and potatoes.

I found amber jewelry in many shops in Leningrad, but the widest collection, including large, beautifully polished pieces, was in the *Berioska,* or dollar shops, where only foreign currency could be used. Russians often lined the counters, even if they had no foreign currency, just to marvel at the radios, hi-fi's, and other electronic appliances. Amber was not cheap,

The Amber Room, in Pushkin (outside Saint Petersburg), before World War II.

and the price increased tenfold if a piece contained an insect. Many shops also sold "ambroid," or pressed amber, which could be recognized by its uniform color and consistency. This product is made in molds by melting small pieces of amber together with a plastic binder under heat and pressure. The final product is still called amber and can be carved into beads or small figurines. Many tourists don't know the difference, but it is obvious when the two are compared side by side.

While browsing in a dollar shop near my hotel one day, I met a tall, distinguished-looking Russian who spoke English quite well. An enthusiastic amber collector, he was obviously delighted to discover an American with similar interests. He asked if I had ever heard of the Amber Room. No? He smiled, took a deep breath, and exclaimed that the Amber Room was the most beautiful and elaborate artistic creation in amber that the world had ever known.

This entire room built of specially selected pieces of amber had been commissioned by the German king Frederick I in 1701, had taken almost ten years for countless craftsmen to make, and was eventually installed in the Main Palace in Berlin. In 1716, Frederick William I, son and successor of Frederick I, signed the Prussian-Russian Alliance with Czar Peter I. To commemorate this occasion, Frederick presented the czar with the Amber Room. It was then installed in the old Winter Palace in Saint Petersburg, but in 1755 was moved to the Catherine Palace in Zarskoje Selo (now Pushkin). During World War II, the German advance on Leningrad was so rapid that the Russians had no time to remove or conceal the Amber Room. Although the Germans never reached Leningrad, whose inhabitants defended themselves gallantly, they did reach Pushkin and there they found the amber treasures. Just as many fine works of art were confiscated by the advancing troops, so was the Amber Room destined to change hands again. The entire room was dismantled and packed in crates, and awaited shipment to Königsberg.

Aware that he now had my complete attention, my new Russian acquaintance continued his story.

The crates were loaded aboard the German ship *Wilhelm Gustloff,* which left the Leningrad harbor at night. But before it could reach Germany the ship was torpedoed and sunk by a Russian submarine. Thus the Amber Room settled to the bottom of the Baltic Sea in no-man's-land. Of course, nothing could be done about it until the war had ended. Several months later, a group of Russian divers found the sunken ship and launched a large underwater expedition to recover the Amber Room. When they finally entered the boat, they discovered that a large hole had been cut into the hull and all of the crates containing the Amber Room were gone.

To this day, the Russian gentleman exclaimed, nobody knows what happened to the Amber Room or where it is now. With that finale, he turned and strode out of the store. I stood still for a few minutes, dumbstruck by his tale, but determined to find out if he was telling the truth.

The city of Pushkin, where the Amber Room was supposed to have been installed, lay only a short distance away from Leningrad, and I coaxed a guide into taking me there. Thus, on April 13, a group of us visited the old Zarskoje Selo, renamed Pushkin shortly after the 1917 revolution because the writer apparently had created some of his works there. The summer palace of Catherine II had been renamed the Children's Village, since it had been converted into holiday camps for the workers. The names of many cities, museums, and palaces were changed after the revolution to reflect a closer relationship to the people (proletariat). Interestingly, I discovered that my Russian acquaintance's story was essentially accurate. We were shown the area that had contained the Amber Room, as well as some other amber treasures (cases, crucifix caskets, chess set) that had been left behind. There was even an old color photograph of the room, the only color photograph ever taken of it.

Since returning from Russia, I have tried to keep abreast of the drama surrounding the disappearance of the Amber Room; and that there are conflicting reports is perhaps only natural. Most accounts claim that instead of being loaded on a boat, the crates were sent by rail to Königsberg, where the room was installed in the main castle. Some say the room was destroyed when the Allies began to bomb Königsberg. But others claim that when the bombing was imminent, the Amber Room was again recrated under the supervision of Gauleiter Erich Koch, the Nazi governor of East Prussia, and shipped to an unknown locality elsewhere in Germany. Herr Koch was imprisoned for war crimes in an old Franciscan monastery in Poland and died there in 1986 without revealing the whereabouts of the Amber Room.

Since the war, this "eighth wonder of the world," valued at $150 million, has been sought after by German, Polish, and Russian agents. In November 1991, Russian president Boris Yeltsin visited Germany and claimed that the Amber Room was hidden in East Germany. Of course, the treasure, if ever found, will be the property of Russia.

Recently, two Germans claimed to know exactly where the Amber Room is hidden. One, Konrad Kirjau, believes that the treasure is in a bunker near the town of Lubbenaw. The other, Hans Stadelmann, is convinced that the crates holding the Amber Room are in a subterranean chamber beneath the town square in Weimar, Germany. Partial investigations at these two sites have not resulted in any discoveries related to the Amber Room, however.

At present, anyone still alive who knows the whereabouts of this art treasure isn't talking, at least not to the press. In the meantime, attention has been directed toward making a replica of the original Amber Room from the color photograph, and the Cabinet Council of the Russian

Republic hired a team of craftsmen in 1979 to begin this enormous task. It will be wonderful to have a replica, but that won't keep most of us from wondering what happened to the original Amber Room.

My three-month stay in Leningrad was nearing its end. This was good in one way because I had lost almost forty pounds, partly due to the difficulty in obtaining an adequate amount of food in a reasonable amount of time. There was so much to see and do; I couldn't afford to spend two to three hours waiting to be served in a restaurant. There were cafeterias, but they were open for only short periods and they attracted long lines of hungry people. So I grabbed food here and there whenever I could.

I spent the final days preparing a lecture to be delivered at the Russian Academy of Sciences—in Russian. I kept thinking of another American who had already presented his lecture and returned to the States. On being asked whether he had any problems with his delivery, he had claimed: "Oh, no, I had no problems. But the Russians, they had a lot of problems." I had never been so nervous before any other presentation. Russian scientists who listened to me practice my talk ended up rewriting portions in their own style. After several of these revisions, my lecture was a confused mess. I finally started over from scratch, this time using simple words that I understood. The delivery must have sounded like a Russian first grader speaking with a thick American accent. But the audience was very kind. They said they understood everything.

Since 1969, another important amber site has been discovered in the northern portion of that huge country. The first expedition left from Moscow in 1970 and headed north to the Taimyr Peninsula in Siberia. The arctic deposits discovered there are twice as old as Baltic amber and extend over a wide area. Thus far some 4,000 insects and other organisms have been gathered, making this one of the largest collections of amber fossils from the period of the dinosaurs (Cretaceous period). The amber deposits in one region are so extensive that a large hill has been called the amber mountain (Yantardak).

After the Russian visit, my Baltic amber fossil collection grew slightly but my collecting mania increased logarithmically. I was learning some of the characteristics of Baltic amber. One telltale sign is the inclusion of stellar-shaped clusters of plant hairs from oak buds. Also, fossils in Baltic amber frequently are covered with a milky white patina. This deposit may be fluid that passed out of the insect into the resin during the embedding process, or it may represent a chemical reaction where the resin meets the insect cuticle. Perhaps it represents bacterial growth. At any rate, it is associated with Baltic amber fossils often to the point of obscuring their structures. Since the pieces with milky deposits are less expensive and therefore more available to collectors with limited funds, my interest in amber from sources that had less expensive fossils with more clarity was beginning to grow. It was time to look at amber from other regions.

3 Looking at Amber Up Close

In the spring of 1973, I attended a lecture in the Department of Paleontology at Berkeley, after which a group of us, including Wyatt Durham and the late Joseph Peck, Jr., began to discuss invertebrate fossils. Knowing of my interest in nematodes, Joe Peck asked me whether I would like to examine some minute fossils in Mexican amber that had been identified as plant fibers but might be nematodes. My interest in amber fossils already whetted, and possessing a basic curiosity about paleontology (sparked by a course taught by Professor Wells at Cornell), I wondered how extensive the fossil record of nematodes might be. Joe took me down to the ground floor of the Earth Sciences Building where the collections were stored. He pulled out the large collection of Mexican amber as well as some interesting samples from other parts of the world. These included less ancient amber from Indonesia and the Philippines, as well as a small collection of Alaskan Cretaceous amber. He showed me one piece of Alaskan amber that contained a biting midge. "Wonder what that was feeding on," he mused. (In 1992, we tried to locate that particular piece, but it apparently had been misplaced). The Mexican amber fossils were all registered on cards, and it didn't take Joe long to find several small pieces that had nematode-like structures in them. I thanked him, and on March 16, 1973, took back to my laboratory a small collection of amber pieces, each neatly stored on microscope slides with labels referring to accession numbers of the Museum of Paleontology. They indeed were fossil nematodes, the first I had ever seen, and I spent the next several years examining them and preparing drawings for what would be my first scientific paper on amber inclusions, or fossils.

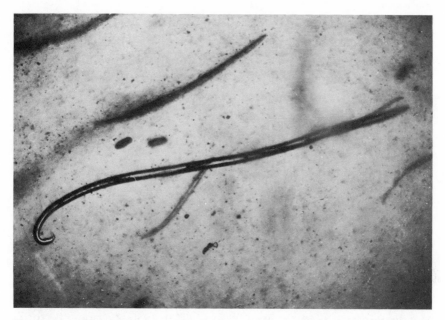

Nematodes in Mexican amber.

I must pay tribute to Joe Peck for bringing these specimens to my attention. Joe, a museum paleontologist at the University of California, Berkeley, from 1949 to 1979, was in charge of the extensive invertebrate collections. He shared his wealth of knowledge with many students, as indicated by the plaque presented him on his retirement, which read "When all else fails, ask Joe." Even after retirement Joe could be found in his office, pondering the possible identity of newly acquired fossils. It was a sad day when we learned, on November 20, 1982, that Joe had died from a mysterious viral infection that he had contracted several weeks before.

Proceeding with my investigation of the fossil nematodes in Mexican amber, I discovered that one block contained not just one nematode but an entire colony, including males, females, eggs, and young of all stages. This reminded me of how the eruption of Mount Vesuvius had preserved humans and other animals in their tracks. Here the sticky resin had flowed quickly over a population of nematodes engaged in their daily events, just as the eruption of Vesuvius had surprised the inhabitants of Pompeii.

Loaves of bread were found under the volcanic ash, and in the amber were strands of fungal mycelium that had served as food for the nematodes. Evidence of injury and disease was found among the dead at Pompeii. In the amber were nematodes whose bodies had been attacked and invaded by pathogenic fungi. The many parallels between the tragedy at Pompeii and the preservation of nematodes fascinated me. Humans don't think of the fossilization of low forms of life as tragic, but to the nematodes it definitely was.

These fossil nematodes of the family Aphelenchoididae held material within their body cavities that seemed to represent traces of the intestine and body wall. The eggs also appeared to contain some cellular material. This rare quality of amber—the ability to preserve not only hard tissue but also inner soft tissue—is one of the phenomena that have always fascinated scientists. As early as 1903, the first scientific paper was written on preserved tissue in Baltic amber insects. Even the title of that paper, by N. Kornilovich, suggests his wonderment: "Has the Structure of Striated Muscle of Insects in Amber Been Preserved?" In 1950, Alexander Petrunkevitch described tissues in Baltic amber spiders. Why wouldn't there be tissues in the nematodes? Roberta, as an electron microscopist, and I took the question one step further. Could the fine structure of that tissue be preserved? In 1976, Roberta began to examine a small portion of the amber piece containing the nematodes for possible cellular remains.

When you have a one-in-a-million fossil, you can't just rush ahead and begin work on it—you might do the wrong thing and lose it forever. So your planning and preliminary research begin with less valuable specimens and pieces of amber without fossil inclusions, to answer as many questions as you can before you even touch your priceless specimen. Fortunately, Roberta was an expert electron microscopist with a wide range of experience. In transmission electron microscopy, fixed and dehydrated specimens are embedded in plastic and sectioned to fine slivers about 800 angstroms thick, mounted on small grids, and then stained to increase contrast before they are examined in the electron microscope. Our fossil was already embedded in amber, so first we needed to determine whether amber could be sectioned like plastic and, if so, how it would survive under the electron beam and in the vacuum of the specimen chamber. We sectioned the first samples with glass knives that we made by taking pieces of glass about one inch square and one-quarter inch thick and breaking them diagonally. The edge at the apex of the two resulting triangles is extremely sharp, particularly in an area about one millimeter from the corner. These knifes worked surprisingly well, producing the thin sections of amber we needed for our study. Not all amber pieces sectioned well, however. Before sectioning, the amber was trimmed to a very small trapezoidal pyramid with a flattened top using a razor blade. We sectioned this surface, which was only about 0.2 millimeters wide. Trimming blocks, as it is called, is precise, tedious work done under a dissecting microscope. The slightest wrong placement of the razor blade can render the block unusable. In our work, different pieces of amber reacted differently to the cutting pressure of the razor—some pieces chipped, others fractured—but we were able to trim some, and these also sectioned well. We could not be sure, however, that we would be able to trim and section the piece of amber with a particular specimen.

One rainy morning the first test specimens were prepared, and we dashed across the wet grass to the nutrition building, where the electron microscope is located. The microscope is tucked away in a half-forgotten room behind some display cases, locked behind two doors and down a short flight of stairs. The room is cool, with a negative room pressure, and is eerily lighted by a yellow safelight and the glowing lights of the operating panel. A gleaming metal column with knobs and protruding handles dominates the center of the microscope, and panels with myriad buttons and dials extend on either side of a tabletop. Midway on the column is the specimen chamber, which is under vacuum and has to be vented to air before the specimen, mounted on a grid, can be inserted into the column. While the air hissed into the chamber, we discussed what might happen to the specimens under the electron beam of the microscope. In transmission-electron-analysis, thin sections, often treated with electron-dense stains, are exposed to an electron beam, and the differences in electron absorption of the tissue components produce the image. Structures range from very electron dense to electron translucent. But the beam is also destructive, and our biggest worry was whether the amber would hold up to the bombardment of the electrons. We had begun with about fifteen samples, and because many had not trimmed or sectioned well, we had only five remaining to take to the microscope.

The first specimen was placed in the specimen chamber, and the pumps hummed as the chamber was evacuated. Roberta slowly introduced the specimen into the column, turned on the electron beam, and then searched for the amber ribbon. Our worst fears were immediately confirmed. As soon as the beam, even at its lowest intensity, hit the sections, it began to burn holes in the amber. Within milliseconds the whole ribbon was laced with hundreds of rapidly enlarging holes, and before we could say "Oh no," the whole ribbon had vanished. The second specimen was destroyed even more dramatically: it literally sizzled and bubbled under the beam. Obviously, this specimen must have had some moisture trapped in it. We quickly replaced it with a third sample. Immediately, black dots rained down all over its surface. The vapors from the previous specimen had contaminated the column, and now the interior of the electron microscope needed to be cleaned. This would take all of the next day. (Electron microscopists have to be patient, philosophical individuals, because the day when everything goes well is rare.) Roberta continued for the next few days working the bugs out of the system, finally obtaining the right circumstances for amber to survive in the beam. During that time, experimenting with thousands of sections and many samples, we found very little to suggest that we would succeed. But we continued, and eventually, to our great excitement, some of the sections that were not destroyed contained obvious nematodes with tissue remains.

These nematodes had thin deposits of tissue still adhering to their body wall, and a few showed a column of tissue running lengthwise through their bodies. We felt that this band could well represent the remains of the alimentary canal. Some of the tissue appeared to have been broken down, and we soon discovered what could have been the cause. Particles identified as bacterial cells were found inside some of the nematode bodies. These bacteria had likely multiplied in the nematode remains, after the nematodes became entrapped in the sticky resin. When Roberta sectioned through one of the nematode eggs, she noticed a distinct layer around the outside of the egg that became detached from the border and could be seen loose around the egg cavity. This was much more definitive than the body wall of the adult nematode that she had previously sectioned through. Because eggs in nematodes possess an extra substance in

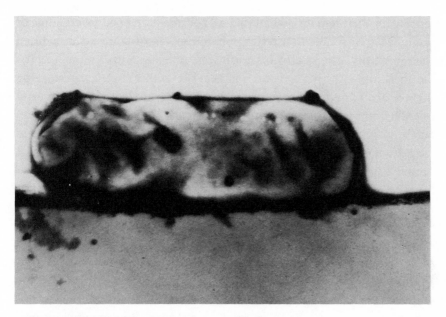

A bacterial cell in Mexican amber.

the eggshell that is absent in other stages, and because this material is more durable than any other component in the nematode's surface, it was clear that we had found the chitinous layer of the eggshell, still well preserved after many millions of years.

The nematode electron microscope project was written up and submitted for publication in October 1976—but the study did not impress the reviewers and the paper was returned. Although the results were not as clear as some would want, we were convinced and, undaunted, decided to continue our search. But now we had to find an organism that would likely demonstrate not only tissue but also definite cellular remains.

The presence of bacterial cells within the body of the fossil nematode raised complex questions. How had the bacteria survived the effects of the amber? Was it an anaerobe, able to survive without oxygen? Had the nematode cuticle kept the resin and associated chemicals from penetrating into the body cavity and allowed the bacteria to thrive on the tissues of the dead nematode? Had the nematodes already been dead, with the bacteria feeding on their remains, when they all became entrapped? Were some of these bacterial spores in a resting phase capable of being revived when their presently adverse circumstances were replaced with a more hospitable environment? This last question kept presenting itself to me, making me wonder. I knew that bacteria had been studied in Baltic amber as early as 1929. In the work of Gustav Blunk, bacterial cells had been obtained after partially dissolving the amber and concentrating the liquid. In fact, if there were bacterial spores in amber, then why not also pollen and fungal spores?

The term *cryptobiosis*—literally, life hidden within a small room—is used to describe the state of life forms when they have shut down their normal metabolic processes (usually in response to adverse environmental circumstances) and have entered a resting/waiting phase. It represents one of the extraordinary means that nature has devised to protect various organisms and ensure their survival.

Most life forms that enter the state of cryptobiosis are microorganisms capable of forming dormant spores. Bacterial spores in canned meat have survived up to 118 years, and examples exist of fungal spores also surviving for extended periods. The ability of most multicellular animals to survive for long periods is very limited. However, rotifers and some nematodes can dry out and remain in cryptobiosis for weeks, months, and even years. In nematodes the resistant stage can be a thick-walled egg or a special modified juvenile stage. Dried juveniles of plant parasitic nematodes have survived for thirty-nine years. A few insects also have this ability. Larvae of a chironomid fly in Africa can survive in the desiccated state for over three years. The seeds of higher plants also may survive for several years. However, the record was set by lotus seeds from an ancient dry lake in Manchuria—they remained dormant for up to 50,000 years and were still successfully germinated. Could some of these resistant spores be revived after remaining embedded in amber for millions of years? It would be four years or so before we would begin experiments to answer that question. In the meantime, opportunities for travel arose, and I was off in pursuit of insects and amber.

4　African Safari

The Balts traded amber with the Romans. The Romans traded amber with the Egyptians, and from there Baltic amber made its way south through most of the African continent, much like the famous Venetian glass trade beads. Most of this amber ended up as jewelry and was frequently sold in local markets held in the centers of large villages and towns. But there was also reputed to be true African amber, which might contain some interesting insect and plant remains revealing ancient African landscapes.

I saw my first African amber in 1976, at an open market in the town of Bouaké in the Ivory Coast. A Canadian government agency associated with the World Health Organization sent me to Africa to study the insect carrying the parasite that has blinded millions of Africans in the Volta River basin. This disease, known as river blindness, is caused by a parasitic worm that lives inside the body of its victims, eventually reaching the eyes and leaving the victims totally blind and dependent on others.

From the standpoint of a parasitologist, Africa can be seen as one large biological laboratory. The more you know, the more you worry and the more neurotic you become. Watch out for that mosquito!—it may be carrying malaria or some pathogenic virus. Don't let that tsetse fly bite you!— there is sleeping sickness in the area. Don't go swimming in that pond!— liver flukes are common here. Look out for the blackflies!—they may have the river-blindness worm. Then there are ticks carrying rickettsias and biting midges with encephalitis viruses. You can imagine how nerve-wracking simply walking from the laboratory to your quarters can be if you attempt to remove each insect that lands on your body.

I remember waiting at the Abidjan airport for a new member of our team who was arriving from Canada. After we drove him through the back roads to his hotel, he refused to leave the automobile. He insisted on rolling up the windows and immediately being driven back to the airport, without ever leaving the safety of the car. He took the next flight back to Toronto and we never heard from him again.

This extreme case of culture shock shows how strenuous the sudden realities of the African scene can be, a point brutally driven home to me during the second week of my stay in the laboratory near Abidjan. One morning our laboratory helpers, who had come from the adjacent country of Upper Volta (now Burkina Faso), failed to appear for work. We waited, and when they had not shown up by the next day, we sent someone to find them. Their bodies were discovered in the underbrush along the roadside; their hearts had been removed for a sacrificial ceremony by members of the local tribe. Panic-stricken, I feared for my own life. Was I going to end up like that? I was assured that only workers from other African states, not visitors from overseas, normally are chosen for such rites. But it showed me how deeply embedded in the culture were the tribal customs. And as if this were not enough, another experience soon increased my tension.

On one especially sultry and still evening, I relaxed on the veranda of our assigned quarters, attempting to identify the unfamiliar sounds emanating from the forest area surrounding our compound—was that a bird, a monkey, a royal antelope, or a pygmy elephant? As I listened, my eyes were fixed on Ajula, an old warrior assigned the challenging task of guarding the compound at night. (Apparently, hiring a local tribesman for this task ensured immunity for the homes in the compound. Every evening Ajula quietly slipped out of the forest and, with his spear and long bow with a few odd arrows, took up his position against a large sloping tree near the center of the compound. He usually appeared to doze off soon after making himself comfortable.)

Suddenly, I saw Ajula quickly move his arm and strike his thigh; perhaps a mosquito, I thought, but I had never seen him move like that before. Ajula then leaped to his feet, knocking over his spear and bow and arrows, and began jumping up and down from one foot to the other, yelling, and slapping his body. Was this some sort of religious rite? Or had the heat and humidity finally broken the resistance of the old warrior?

He kept screaming something that sounded like "maniok." I yelled to Adama, the houseboy, who was washing the dinner dishes inside. He ran over, took one look at Ajula, and exclaimed, "Army ants! It's an invasion of army ants." Immediately I felt a stabbing shock in my right calf. I slapped my hand against my leg and looked down to see a large struggling ant at my feet. "A scout," said Adama behind me. "That means they will soon be here."

At that moment a mélange of small creatures flew, ran, or jumped up on the veranda—katydids, roaches, beetles, lizards, and toads—all trying to escape the marching death. Under these circumstances their normal predator-prey instincts were on hold; lizards abutted against tasty beetles without giving them a second look.

"Adama, what should we do?" I asked. I had heard of army ants entering houses and attacking everything in their paths, including pets and even infants. Some of the houses behind mine contained children now asleep in their beds.

Adama ran to the kitchen and returned with some chicken scraps from the dinner meal. "This may work," he panted. He ran down the steps and tossed the remains toward the woods. I cautiously descended the stairs to see what would happen. The chicken skins suddenly disappeared against the darkness as thousands of ants engulfed them.

Suddenly, I felt points of pain on my legs. My God, the ants were attacking me! I had always considered myself a calm, rational person, but instinct took over and I found myself jumping around, slapping my legs, just as Ajula had done.

Obviously, the chicken-scrap plan wasn't working, and I looked around anxiously for Adama. Why had he deserted me now? Then I saw him. Out in front of the advancing tide of ants, just twenty feet away from the veranda, he was pouring something on the ground, constantly brushing ants from his legs. Then he stopped, took a match from his pocket, lit it on his heel, and threw it on the ground. A semicircle of flame rose up, and a few moments later I heard the crackling of ant bodies and smelled the roasting of insect cuticle. But the ants kept advancing and their bodies smothered the flames. Could nothing halt them? Adama poured another barrier, and I saw with alarm that the kerosene can was nearly empty. Would the ants smother this one, too? Adama stood by my side, and we

Open market in the Ivory Coast where "amber" beads are sold.

watched the flames as they continued to burn. We could see masses of ants behind the barrier, and then suddenly the black tide turned and began to move back toward the forest. Had we won? We held our breaths. The ants were disappearing into the darkness. I embraced Adama and thanked him for his quick thinking. The experience gave me a healthy respect for those small creatures that could provoke great fear in all other animals, large or small, including humans.

To ease the tension of these unaccustomed experiences, I sometimes took off and strolled for hours through the markets, discovering fascinating sculptures and miniature gold weights, but always on the lookout for amber. There were some small pieces of amber in Bouaké, but the markets in Bobo Dioulasso in the Upper Volta had much more. I convinced some Canadian acquaintances in Bouaké to drive up there with me. We proceeded north on washboard dirt roads all the way, passing through Senufo country in the north of the Ivory Coast. The Senufo buildings were characteristic, round mud huts with small rectangular windows topped with grass thatching. The grain-storage huts were like miniature homes, built the same, but smaller. We passed donkeys patiently pulling carts along the road and came across herds of zebu cattle guarded by tall, slender people from the north.

Near the border with Upper Volta we passed over the Leraba River and watched the splashing hippos. A family of monkeys crossed in front of us, and flocks of quail-like birds scurried through the undergrowth.

The border guard didn't seem to care that we had no visas. I gave him my pen and he waved us onward to Banfora. Just after we left this village, a cloudburst halted us in our tracks. What had been a roadbed minutes before suddenly became a wide, rushing river. We waited until some of the water subsided, and then we continued. By now darkness was descending, and we saw many Africans walking along the road with rifles over their shoulders. We asked one if he was hunting. No, he exclaimed, the arms were for protection against panthers.

We finally arrived in Bobo after a full day's drive and were allowed to stay in the guest house of the Veterinary Science Laboratory, well known for its program to combat the dreaded tsetse fly. Scientists from France and Belgium had developed a technique for irradiating and sterilizing the male flies, then marking them with a small dot of paint and releasing them in wooded areas along streams. When the treated males mated with naturally occurring females, the eggs would be infertile. There was much excitement in the laboratory, because the first field experiment using this method to eradicate the tsetse fly in isolated areas was being planned.

That night, tired as we were, we decided to go to the one and only movie theater in Bobo Dioulasso. Months had passed since any of us had been to the cinema, and television didn't exist in the areas where we worked. Two films were being shown: a Spanish movie on Dracula and an Italian western. We were advised to pay a bit more and sit in the first-class section. After a half-hour delay, the lights went out and the Dracula film started. It was a blood-curdling thriller but difficult to follow because it was in Spanish and people constantly walked back and forth in front of the projector lens. In addition, several heavy thunder showers pounded down on the metal theater roof, drowning out all dialogue. Several unscheduled intermissions occurred when the film broke and had to be repaired. During the romantic scenes, the projectionist used his discretion to censor what he must have considered the "X-rated" portions, either by increasing the brightness to blind the audience or by adjusting the focus so we could see only a blurry Dracula attacking his hazy victims. During the Italian western film, which was much better received by the audience, the film shifted to one side of the screen and we watched ten or fifteen minutes of film perforations along with the left side of the pictures.

That evening, when we returned to our quarters, we noticed the cook running this way and that way with a bucket of water. He would suddenly drop to all fours, pick up something from the ground, and throw it into his bucket. He explained breathlessly that the recent downpours had brought out large swarms of huge flying termites, which were available only several times a year. When stripped of their wings and boiled for several minutes, they were supposed to be delicious. He insisted we come into the kitchen and wait while he prepared a batch. The boiling took only a few minutes, during which we all dared each other to eat them. But when we saw the large brown bodies and tan-colored wings floating on top of the oily water, we suddenly lost our appetites and decided to leave—but not before the cook laughingly placed several of the greasy cadavers into his mouth.

We spent the following weekend in the market looking for amber, which certainly was more plentiful here than in the Ivory Coast. With the exception of several small pieces, this African amber was yellow and opaque. I purchased several small pieces of all the different types since I had heard that fake material was common. Late at night I examined and tested them and discovered to my dismay that all, except for two beads that were shaped differently from the rest, were made of plastic. The most common type of plastic beads were tubular, barrel-shaped pieces varying from half an inch to one-and-a-half inches in length. The merchants told me that it was Somali amber, but I remember seeing similar material in Italy and even in the shops in Berkeley. I began to wonder if any naturally occurring amber existed in Africa. I studied the two true amber pieces in detail and the following day returned to the market to search for more. The real amber was difficult to find, and often was beaded on a long necklace together with a majority of plastic pieces; the only way to obtain it was to purchase the entire necklace, which wasn't cheap.

Answers varied as to where the real amber originated. Everyone agreed that it came from the north, but where? Several merchants said that it was Berber amber, which naturally occurred in the Atlas Mountains in Morocco. Perhaps I should go there.

In the meantime I continued to buy as much amber as I could. Obtaining merchandise in an African market is not a clear-cut transaction. You never pay the asked price (unless you have just stepped off the plane), but engage in an artful discussion of the price using the process of bargaining. This old and standard method of obtaining goods in many countries is certainly part of the African cultural tradition.

Bargaining is a type of social communication that is carried out to only a limited extent in the Western world. How many of us would walk into a department store, take a shirt up to the counter, and tell the clerk that we are willing to offer half the asking price? But this is common in Africa. Bargaining is something like a verbal wrestling match in slow motion. The final price depends on many factors, including the feelings each party has for the other, the financial state of the parties (whether the seller is deeply in debt or the buyer wealthy), the weather (cheaper just before an approaching storm), time of day (cheaper in the late afternoon just before the market closes), time of year (the arrival of some merchandise is seasonal), and the amount of time you want to spend. Most Americans don't like to bargain; it takes too much time and is too personal, and some feel it takes advantage of the seller. But I have spent as long as two hours bargaining for an amber necklace. Most African merchants expect and enjoy bargaining. If you take longer than ten minutes to bargain for any single object, however, they will be very upset if you don't eventually buy the item. Many times the final price was shouted out after I had begun to walk away. Later I learned when to walk away and when to stop bargaining. One important point to remember: you will be held for any amount you utter, so be careful what you say.

Because of the nature of our work, we often had to spend the night in the bush. Again, the biggest problem was not marauding cats or canines but insects, especially mosquitoes. We normally carried one-person hammock tents, essentially a hammock with a canvas roof and mosquito-netting sides. We strung it between two trees and climbed in quickly, to avoid catching mosquitoes inside. I woke up one morning and realized that during the night I had rolled over and laid my arm against the mosquito netting, which left me with rows of mosquito bites accordingly. I must have fed nearly twenty-five mosquitoes during the night, and it turned out that at least one was infected with the malarial organism.

The fever came on slowly, the first symptoms being nausea and malaise. Adama prepared some citron tea from the lemon grass that grew behind the house, hoping that it would make me feel better. After Adama went home to his family, I began to feel cold; how strange to sit nearly on the equator at sea level and feel cold. I reached for the tea cup and noticed my hand shaking—I was shivering. The cold became almost unbearable. I ran into the bathroom and turned on the hot shower faucet as far as it would go and climbed underneath. I couldn't imagine removing my shirt and shorts—I was too cold. I stood there shivering, with the steaming hot water running over my clothes. I began to panic when I noticed the hot water getting cooler. But just then I was overtaken with intense heat. The room started to spin. I shut off the hot water and opened the cold, obtaining the same feeling of relief from it as I had from the hot water a minute or two before. After what seemed like an eternity, the fever broke and I fell, exhausted, into my bed. I thought that I might be close to dying in this isolated town in the middle of West Africa far from the people I loved.

The next morning, having survived the night but still trembling, I drove my Jeep to the Treckville Hospital, where the doctor remarked that the malaria in this area was resistant to the normal medication dose and so the amount must be doubled. Double it I did, and I returned to normal after a few days. But long after returning to the States, if I stressed myself with overwork and lack of sleep, the chills and fever returned.

Just before I left West Africa, Jan, a Dutch scientist temporarily on leave with his family in the Ivory Coast, invited me to go sailing. (I had always loved to sail and had been a sailing instructor at the University Yacht Club in Berkeley.) The boats were located at the inland tip of the lagoon near the village of Adiopodoumé. The lagoon, a wide expanse of brackish water, connected directly with the open ocean. The yacht club that Jan belonged to consisted almost exclusively of visiting Europeans and their families. The African-built boat that we took, about twelve feet long, belonged to a class I had never seen. The single mast looked a bit thin, but I didn't expect any problem. We were quickly propelled out into the middle of the lagoon by the gusty winds. When Jan asked me to take the tiller, I noticed that it looked weak and had no catch.

We had now sailed past the middle of the lagoon, and the open sea loomed ahead. Time to tack and head back, motioned Jan. I pushed the tiller—and heard a sharp splitting sound; the handle had broken, spinning us around so sharply that the wind caught us broadside and cracked the mast. Over we went into the warm water of the Abidjan Lagoon. I had capsized before, but never had broken a mast or tiller. Aside from the discomfort and loss of a few items from the boat, it normally isn't a serious problem, especially in warm water. I motioned to Jan to take the bow and I swam to the stern. We soon learned that righting a boat with no air tanks was impossible. We first attempted to lift it out of the water, but with both of us lifting, there were no hands to bail; the boat sank again beneath the water. I anxiously noticed that the water was only slightly brackish at this point and immediately imagined fluke larvae penetrating my skin. But we could not right the boat, and worse than that, we were rapidly drifting toward the open sea. Once we were swept out there, sharks would certainly finish us off. I had watched them swimming along the shore several weeks ago—those gray fins skimming between the waves, picking up fish that had been injured during the daily fishnetting activities.

We yelled, but by now were too far from any shore. In fact, I couldn't even see the small white houses that marked the yacht club. Besides, we were exhausted and could only hold on to the half-submerged boat. Now I wished that I had thoroughly inspected the boat. With all the prevalent diseases I was constantly exposed to during my stay, I never imagined I would end up as shark food. We were nearly in the mouth of the lagoon, and small waves washed over us as we clung desperately to the half-exposed hull of our capsized vessel. Suddenly, I heard what sounded like a motorboat. Jan pointed feebly over my shoulder. I turned around and spotted an old, weather-beaten motorboat heading in our direction—what a beautiful sight. The Frenchman threw us a line and we ran it through the bow line on the sailboat, gave it back to him, and half crawled and half fell into his boat. Back at the yacht club, I asked our rescuer how he had known we were there. Pointing to a small speck in the water, he said: "That African fisherman in the dugout canoe saw you and came to tell me. I would have never known. They have good eyesight, you know."

I was relieved to return to Berkeley. On testing all the amber purchased in West Africa, I found that about three-quarters was plastic, a large percentage of the real amber originated from the Baltic area, and a few pieces were copal from East Africa. I concluded that the only real African amber must be Moroccan amber from the Atlas Mountains.

5   Adventures with Tomb Amber

Baltic amber was traded extensively throughout Europe, well before the birth of Christ. There have been numerous finds of Baltic amber in tombs dating from the Bronze Age and Iron Age. One of the most famous was that of Heinrich Schliemann in 1876, when he discovered large quantities of amber beads from necklaces in the prehistoric Greek city of Mycenae. The natural question was, where did this amber originate? In 1963 the organic chemist Curt Beck began using infrared spectra of European tomb amber to determine whether the samples originated from the Baltic. All Baltic amber produces infrared spectra with a characteristic pattern distinguishable from the spectra of other European amber. Beck's results indicated that the majority of amber beads from the shaft graves at Mycenae did originate from the Baltic region, thus providing evidence of trade between these regions from 1450 to 1100 B.C.

Archaeological amber sometimes turns up in strange places. In the National Museum in Copenhagen, some 4,000 amber beads from Stone Age farmers (4200–2800 B.C.) are on display. These were recovered from bogs in Jutland, buried there by farmers in hopes of obtaining prosperity for their crops and livestock. Curiously, these beads were accidentally discovered during World War II, when people dug up and used bog material for fuel. Also displayed in the National Museum are amber carvings of game animals that served as amulets. These sculptures, made by Stone Age hunters (7500–4200 B.C.), represent some of the earliest carved amber figures. Still earlier, and carved in the shape of a wedge, is an amber piece dating from 8600 B.C. This might have represented an axe, since an amber

33

Amber beads from a Bronze Age tomb in southern France (500 B.C.).

axe combined the magical powers of both amber and the axe, supposedly making the bearer invincible.

Other Bronze Age and Iron Age amber artifacts have been reported in Croatia, Bosnia-Herzegovina, and Italy. In the latter country, especially beautiful amber jewelry was made by the Villanovans, an early Iron Age culture that lived near Bologna and formed part of the Etruscan Empire. Amber ornaments in tombs from 750 to 650 B.C. consisted of necklaces and pins made with copper. The Villanovans were accomplished metalworkers and controlled the copper and iron mines in Tuscany.

There are few museums in major European cities that don't have some display of prehistoric amber jewelry. Amber ornaments of all types have been recovered from graves, dating from the so-called Celtic period (1000 B.C.–A.D. 600). An especially well preserved amber bead on a bronze ring is displayed at the Swiss National Museum in Zurich. Thus I was eager to visit some prehistoric tombs and examine some of this amber firsthand.

In 1978, while working on parasitic nematodes (similar to the parasites that cause river blindness), in a biological laboratory in Antibes in

*A bronze ring with amber
bead from the Celtic period
in Switzerland (1000 B.C.–
A.D. 600).*
(The Swiss National Museum,
Zurich)

*Portion of an inscribed
amber disk from an Iron Age
tomb in Croatia.*
(Specimen provided by Joan
Todd)

southern France, I read of an archaeological dig near the city of Toulouse. I convinced my colleague Christian Laumond to call the archaeologist and ask if we could visit the site. We obtained permission and the next day drove off to what would be my first excavation of archaeological amber.

The archaeologist, Bernard Pajot, explained that the area was a burial site for the Celts and dated from 500 B.C., or during the early Iron Age. Apparently, the bodies of the deceased were burned, and then the ashes and bones were collected and scattered over a small area of about twenty-five square feet. Various ornaments and weapons (sculptures, buttons, knives, daggers), including amber beads, were placed over the human remains, and then the area was covered with soil and stones. The amber beads were round and flat, each with a hole through the center. At first I didn't recognize the beads as amber, because their surfaces were badly oxidized and covered with soil. But their light weight differentiated them from stone or clay beads.

Witnessing my excitement over these 2,500-year-old amber beads, the archaeologist asked if we would like to visit a second site, this time a shaft grave dating from the late Stone Age (1800 B.C.), and look for similar beads. "Why not?" we replied. In this culture, the people searched for a large stone, called a Dolman, and erected it at the entrance to a tomb, the latter being a ten-foot-square shaft that went into the earth some twenty feet. The deceased were wrapped in cloth or leaves and simply dropped in the hole. Bernard had spent two years removing the soil from that tomb and lining the sides with a wood frame that was used as a ladder for entering the grave.

The air became humid as we descended, and the muddy soil at the bottom, mixed with human remains, created a unique odor. I heard a board begin to crack. I looked up and saw Christian, a large man, holding onto one of the uprights. But his 200-pound bulk was too much for the beam and, with a loud crack, it broke. Christian went hurtling down, striking Bernard and carrying them both to the bottom of the pit. I looked down at their still bodies, lying among the bones of their ancestors. My first thought was "I hope they're still alive." My second thought was "How in the world will I ever get them out of this tomb if they're badly hurt?" Bernard was the first to move. In breaking Christian's fall, his head had struck one of the beams and was bleeding. I helped him to a sitting position. Christian moaned, and I gently raised his face out of the mud; he had broken his arm. After about fifteen minutes, I helped my companions crawl out and drove them to the local hospital. We never did have a chance to look for amber in that tomb.

Although tomb amber is usually highly oxidized, the thick crust that covers it can be removed with sandpaper. The matrix may still be solid and even composed of clear amber, although it usually darkens and turns reddish from exposure. Even biological inclusions may occur in tomb amber, just as they can turn up in amber necklaces made today. Of special interest to archaeologists are inscriptions or designs carved on the surface of some amber pieces found in tombs.

6   Excursion in Poland

During my stay in France, I visited some amber areas in Poland. One of the most interesting and historically complex countries in Eastern Europe, Poland is unfortunately located between two powerful nations (Germany and Russia) that often fought their battles on Polish soil. The conflicts that Poland has experienced have been reflected by its expanding and contracting borders throughout history. The Poles like to point out that there once was a Great Poland, with an eastern border that reached far beyond Kiev, a northern border that reached beyond Riga, and southern and western territories that included most of present-day Slovakia and part of southwestern Germany. But that configuration did not last long. The history of Poland is filled with battles and destruction, starting long before the Mongol invasion that destroyed the beautiful city of Kraków in the 1200s.

Poles regard most native-born Americans as naive souls who don't know what suffering is, have never lost their homes in a war, and have never experienced a war on their own land. Because of this they feel that most Americans can never understand the Polish soul. Had it not been for the determination and stubbornness of the Poles, their country might have been swallowed up by its surrounding ambitious neighbors. As a Pole in Warsaw warned, "Be careful, life is brutal and full of traps."

I was not aware of this sentiment during my first visit to Poland in 1978. I had been invited to participate in an International Congress of Parasitology held in Warsaw. It was raining hard when the plane landed, so I took a bus to the Hotel Metropol in Warsaw only to discover that my "reserved" room had already been taken and the hotel was full. In contrast to

Russia, Poland placed few restrictions on its people, and Poles were allowed to put up foreign visitors in their homes. At first I was hesitant, but after several attempts to locate the Palace of Culture, where I had been told to go to make another hotel reservation, I accepted a room from a Pole standing outside in the rain holding up a sign, "Room, Zimmer, Chambre." This was one of those times when you shrug your shoulders and trust to your fate.

The fellow drove to his flat, which was located in a large rectangular apartment building, along with forty or fifty other buildings all lined up, on the outskirts of Warsaw. He introduced me to his wife, a Warsaw native who worked full-time. Both spoke some English. They owned a forty-pound Newfoundland puppy—their pride and joy—and, given the small bath, small kitchen, and combined living and bedroom, I looked around anxiously to see where everyone would sleep. The man pulled a folded air mattress out of the small closet and inflated it. Thus my first night's sleep in Poland was interrupted by a Newfoundland dog pulling off my covers and gnawing on my feet as I lay helpless on a partially deflated air mattress on the floor. I remember thinking that perhaps a second glass of vodka, which had been offered earlier, would have been a good idea after all.

The following day I found a room with a bed and no dog, and remained there during the rest of my stay in Warsaw. I paid only $5 a day, although the normal rate for Americans was $25 a day; for Poles it was only $1 a day.

One important attraction for amberphiles in Warsaw is the Earth Museum, which has a permanent amber exhibit emphasizing the cultural and historical aspects of Polish amber. Although amber can be dug from the ground in many parts of Poland, most deposits are along the Baltic coast, where they are washed up on the beach, especially after heavy storms when the turbulent wave action moves the larger pieces.

Aside from picking up the amber by hand, the simplest and most ancient method of amber collecting is the use of what appears as a long-handled butterfly net that can be shoved eight to ten feet into the surf to snag loose amber pieces. Even longer-handled nets can be used to recover amber in shallow water with a rowboat.

In the eighteenth century, open pits were dug in the amber-rich sand bordering the Baltic Sea, as well as in the soil of certain forests. As the pit

Netting amber along the coast of the Baltic Sea.

filled with water, lumps of amber fell out of the soil and collected in the bottom of the pit. Floods are one means by which the amber was washed southward from the Baltic Sea. Still further movement of amber from the original site can be attributed to the action of glaciers that carried the amber in the ice pack. It is not unusual to discover large pieces of Baltic amber with deep parallel ridges cut into the surface by glacial action.

Although the greatest demand for amber was for ornaments, it was, and still is, used in folk medicine and for amulets. Small, sharply pointed amber "icicles" were used to excise items from the eye; alcohol containing partially dissolved amber particles was taken to treat lung ailments; amber powder was made into salve for the treatment of skin disorders and was sometimes used as snuff to treat the flu. Amber teething rings were given to infants; not only did the cool touch soothe tender gums, but the amber also was thought to inhibit bacterial infections.

Protection from evil forces, and luck in hunting and mate finding, were the properties that amber amulets were supposed to endow. Thus hunters

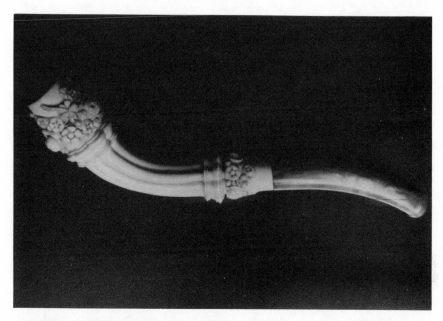

A meerschaum pipe with amber stem.
(Museum of the Earth, Warsaw)

during the tenth through thirteenth centuries A.D. wore or carried with them on expeditions small amber carvings, such as a hatchet, a knife, or a miniature of the animal being sought. Other good-luck charms appeared in the shape of the sun, or were spindle shaped to assist weavers in keeping the yarn straight. During Roman times, the emperor Nero is said to have allowed the gladiators to wear amber amulets for protection and victory. Amber pieces decorated the Coliseum.

Amber charms also were carried by sailors to provide a safe passage. They didn't always work, of course; in a Stockholm museum lies a piece of rough amber removed from the captain's cabin of the flagship *Wasa,* which sank in the Swedish harbor in 1628. Even today the Bedouins in Israel wear amber necklaces for good luck; and not only do members of the Santhal tribe in Bengal themselves wear amber for protection, but they place amber necklaces on their cows to prevent mishaps.

Still other items made from amber were stems for pipes, cigars, and cigarettes. Amber supposedly imparted a "cool" taste to the smoke and was pleasant to hold. Such stems, often combined with meerschaum, first appeared in the eighteenth century and are still made today, although many modern ones are pressed amber and possess a higher percentage of plastic than amber.

The eastern amber trade route began at the Baltic Sea in Poland and eventually reached Greece and Rome. Caches or hoards of amber objects together with Greek and Roman coins mark the routes as they transected the Polish landscape. There is evidence that trade with the Romans was quite lucrative for the Balts. Graves in the region of Königsberg (now Kaliningrad in Russia) revealed skeletons in which Roman coins still lay between the teeth of the deceased. After all, if the dead wanted their souls to cross the river Styx into the underworld, they had to pay the boatman Charon, and what could be more acceptable than a Roman sestertius? Most recovered coins date from A.D. 138 to 180, during the reign of Emperors Commodus and Septimius Severus. The amber trade with Rome came to a sudden halt after the collapse of that empire in A.D. 475, and it was not until the tenth and eleventh centuries that a trade agreement was reestablished, this time with Egypt, Arabia, and other Near East countries. That the trading was carried out through Arab intermediaries is certain, thanks to finds of silver dirhem in amber caches dating from this period.

The revitalization of the eastern amber trade route was accomplished by the Teutonic Knights. These Knights belonged to a religious group that had returned from the Crusades in the 1100s and settled in an area along the Baltic Sea then called Samland, or Sambia. They replaced the original Prussians, whose ancestors had collected and traded amber for centuries, with a well-organized military rule. The Knights declared their rights over all of the amber washed up on that part of the Baltic coast, thereby establishing a strict monopoly. Those who did not comply and were found collecting amber could pay with their lives. Many local citizens attempted some organized resistance. But the Teutonic Knights were too strong and well established, and they split the opposition by offering immunity to certain individuals or groups, such as the Bishop of Sambia in 1274, the fishermen of Gdansk in 1312, and members of a monastery in 1342. Almost everyone else, including all of the amber craftsmen in the area, executed their work only on command from the Grand Master of the Order. Control by the Teutonic Knights began to weaken around 1400, however, as military commitments began to drain their time and power. Finally, after their defeat in the battle of Grunwald in 1410 by the Poles and Lithuanians, the amber market was free. But not for long. Almost immediately afterward, a number of competing guilds sprang up throughout Poland and Germany, and effectively gained control of the amber trade.

The strong tradition of amber carving had its center at Gdansk, where the first artistic amber guild was established in 1480, to be followed later by other centers at Elblag and Slupsk. The imagination of these amber craftsmen was truly phenomenal, and sculptures of saints, historical figures, and animals adorned caskets, crucifixes, and other items. Unfortunately, many of these works of art were destroyed during the last world war, but some remain in various museums throughout the world.

Many examples of Polish folk-art amber carvings adorn the display cases in the Earth Museum. These consist of necklaces composed of flat round beads, brooches, medallions, and finger rings, and are especially abundant from eighteenth- and nineteenth-century excavations from the Koszuby and Kurpie districts in Poland. All the early craftsmanship was completed by hand and took many long, tedious hours of work. In the eighteenth century, spinning wheels were converted into primitive amber lathes and grinders, and lamp glass was used in place of modern-day sandpaper.

Barbara Kosmowska-Ceranowicz, head of the amber portion of the Museum of the Earth, was a wonderful guide, and in addition to taking me through the permanent display, she showed me rare scientific and historic pieces that are normally kept behind locked doors. Piotr Mierzejewski, a shy, quiet Pole, was using the scanning electron microscope to examine the surface of preserved insects that had been dated as 40 million years old. He was particularly pleased with the discovery of well-preserved book lungs in the body of a spider. We discussed our common interest in the preservation of amber-embedded organisms. They showed me long wooden sticks with a series of backward-pointing hooks at the tip that the searchers pushed into sandy soil to detect amber. They could tell by the vibration whether the stick had hit a stone or a large lump of amber.

On leaving Warsaw and its interesting mix of people, I was driven by a Polish scientist north to the town of Gdansk (Danzig, when the territory was occupied by the Germans) near the Baltic coast. It was in this territory that the Pruski peoples (original Prussians) lived some hundreds of years ago. The whole area reflects a history of conflict and war—the Pruski people were eventually annihilated as a distinct entity in the seventeenth century.

The old town of Gdansk is a delight and seems almost a copy of an old Flemish or Dutch town with narrow gabled houses. Saint Mary's Street was lined with amber artisans, and I could have spent weeks exploring each shop and examining all the varied amber articles. I met Zdzislaw Kycler, a modern amber sculptor who had just finished carving a religious scene from a huge piece of amber. It had been purchased by the Catholic church and was destined for the Vatican Museum.

I wanted to see the old mines in Baltijsk, across the border in Russia, but no one admitted to ever going there. Later some Poles told me that night boats frequently went there to smuggle amber back into Poland. "Business knows no political boundaries," they told me laughingly.

But the Poles were eager to discuss the fate of the Amber Room. They confirmed that it had been removed from Russia and placed on the *Wilhelm Gustloff,* a German transport boat operating toward the end of the war. The boat was then torpedoed near Koszalin in fifty-seven meters of water with 6,000 men aboard. A team of Polish divers had explored the sunken ship, discovered that a hole had been cut through the hull, and found that the boxes containing the Amber Room were gone.

A number of shops were selling "black amber," which was solid black and shiny. They insisted that it was a new type of amber. On returning to the States, I tested it and discovered the product to be jet, a highly compressed form of wood or coal, and not amber.

Much of the amber jewelry in the Gdansk shops contained yellow to brown colored disks in the amber matrix. Shop owners in the States had explained that they were "fish scales" or "sun spangles." They certainly didn't look like fish scales under the microscope; besides, how could scales from an aquatic animal ever make their way into amber—unless the animal was a tree-climbing fish? The puzzle was solved in a factory where they make these disks in clear amber. The exact process was kept secret, but it involves heating the amber in sand for dark scales or in oil for light scales. These disks or scales are internal fracture lines, and if the heating process is too fast or too long, a pile of fractured amber pieces is the result.

These same factories were also making ambroid or pressed amber. They took the small pieces left over from previous operations and pressed them together under 100 atmospheres of pressure at a temperature of 120 degrees Celsius.

Another treatment conducted here was the clearing of amber. Although some find the cloudy or opaque amber by itself quite attractive, higher prices are paid for the clear material, and, of course, the cloudy material masks any inclusions. Cloudy amber is made clear by heating it in various types of oils that slowly enter the amber matrix and fill the tiny air bubbles that give amber its cloudy appearance. Amber was cleared in historic times as well, commonly in an iron pot filled with rapeseed oil over a kitchen fire.

My Polish scientist friend and I then left Gdansk for the Baltic coast and searched the sands for amber. Small pieces weren't difficult to find. In fact, many tourist food stands that lined the beach were selling necklaces made from pieces of amber picked up directly from the beach. A small hole was drilled through the center of the pieces, and they were then beaded on a piece of ordinary string. Farther down the beach we met a fisherman returning with his daily catch. He invited us to his home, a small, cozy stone cottage nestled among the trees in a small woods about a kilometer from the sea. Pots of fish soup steamed on an iron wood-fueled stove, and an old table near the window contained an assortment of amber he had picked up on the beach. The amber from the Baltic Sea was generally more varied in color, shape, and size than what I had encountered along the North Sea. My eye fell immediately on a beautiful piece of yellow cloudy amber that was partly polished and yet still showed deep gouges on one side, probably from glacial action. There were many other irregular-sized pieces, most exhibiting a reddish crust on their surfaces from their long-term exposure to the air. But my eye kept returning to that beautiful golden piece. I had never seen anything like it. After consuming several bowls of fish soup and some Polish vodka, I left the fisherman's cottage with the piece in my pocket, and the fisherman's family was $200 for the better. Little did I know what I would have to endure to keep this piece.

On the way back to Warsaw, we stopped at the famous Castle at Malbork. This impressive fortress, with its moat, redbrick fortified walk, and watchtowers, had been one of the last strongholds of the Teutonic Knights.

Ornate amber box made by Christopha Mouchera in the seventeenth century.
(Malbork Castle Museum)

A showplace of amber craftsmanship, it contained exhibit cases of six-teenth- and seventeenth-century amber caskets, crucifixes, and other beautiful works of art.

We couldn't stay long because I was almost late for my train to Berlin, so we hurried back to Warsaw. I thanked my friend for taking me on this wonderful journey, gave him money for gas and food, and waved goodbye as he sped away.

At the station the customs agent asked that my bags be opened and then examined them thoroughly. I thought I detected a note of despair when he didn't discover anything important. He then turned to me and searched my coat. It didn't take him long to discover my prize piece of amber. When a wry grin spread across his face, clearly saying "I have you now," I knew that I was in trouble.

In such situations, feigning ignorance is sometimes the best policy. But you can't play dumb if you don't know what it is you're not supposed

◄

Built in the thirteenth century, the Malbork Castle in northern Poland served as the base from where the Teutonic Knights controlled the Baltic coast amber trade.

to know. The agent whistled, and a short, stout man appeared. From the number of medals and patches on his uniform, it was obvious that he was the chief customs official. "It is forbidden to take unworked amber out of Poland," he said. I was then whisked into a small side room and ordered to remove all my clothing down to my underclothes. Seeing no reason for this, I objected. Besides, my train was about to leave. I suddenly remembered the stories I had heard in Warsaw of foreign tourists being locked up for weeks and months in Polish jails for some small offense. How large a crime was it to attempt to leave the country with a piece of amber? By now the inspector had my wallet and was removing all my money. I objected again, to no avail, but luckily the money in my wallet was apparently enough. They detained me another thirty seconds, and then the head official grabbed my clothes, threw them at me, and said, "Go now." I did as he said, happy to get my freedom back regardless of the price. As I threw on my coat and grabbed my suitcase, the official yelled. Oh no, what now? The train was moving away from the station. I turned to look at him. With a triumphant little smile, he picked up the piece of amber and threw it to me. I turned and ran, jumping into the still-open door of the last car. It took several minutes for my adrenaline level to drop; then I looked for my seat.

I finally found my compartment in a forward coach. The other five passengers already seated there looked up to greet me, and then watched curiously as I wedged the piece of amber under my shirt, hoping to avoid any further trouble.

Waves of customs officials came through our small compartment, and I worried that they would find and confiscate my now extremely expensive amber souvenir. Each time, they asked one of us to open our bags, and by the time they had finished we all knew what each suitcase contained.

The most traumatic experience occurred near the border between East and West Germany. A soldier entered and looked under our seats with a flashlight, making us all hold up our legs so he could get a clear view. Meanwhile, another soldier was directing guard dogs under the train in a search for anyone who might be hidden there. Then we heard footsteps on the roof as another guard patrolled overhead. An old East German in our compartment leaned forward and whispered: "They are looking for deserters. In West Germany, there is a museum that shows how people have escaped. One man climbed in a cement mixer and remained there for five days before making it across."

Another guard entered our compartment. He turned to an American student and said, "You can get out here and catch an East German train to Denmark." The student objected. He had a ticket from West Berlin to Denmark and he did not want to take another train. The guard continued, "Get your bags and come with me." At that point the frustration we all felt with the bureaucracy boiled over. We burst forth one after the other, exclaiming that this poor soul should not be made to take a train he didn't want to take. Suddenly, we grew quiet, staring at one another almost in disbelief. What had we done? We were still in East Germany. They could lock us all up. But the soldier shrugged his shoulders, turned, and left the compartment. We had stood together against the system and apparently gotten away with it. For the rest of the journey we were filled with camaraderie.

When the train finally reached West Berlin, we felt we had reached the Promised Land. It was wonderful to feel free again. My only sadness was in leaving these new friends, who for a short time had seemed almost as close to me as members of my own family. But at least I did still have my beautiful piece of Baltic Sea amber.

On January 29, 1993, Poland issued a set of postage stamps paying tribute to the rich history of amber. One large stamp commemorates the ancient amber trade route by displaying a Roman map of western Europe with the amber route marked with a series of amber beads. In the lower corner of the commemorative envelope is a group of Roman coins and an oxen cart laden with its precious amber cargo.

Present-day amber dealers who have recently returned from Poland report that the collecting of amber today resembles in some ways that of the past. Organized groups have now claimed rights to amber found along the Baltic coastline, and those who do not heed "regulations" have been found floating in the sea. Paying for amber with one's life has occurred time and time again during the past 2,000 years, and possibly even longer.

I returned to the laboratory in Antibes, where I was to finish my sabbatical leave. Morocco was not far away, and if I would ever find real African amber, this might be my only chance.

7 Searching for Berber Amber

Finally, in 1979, I arranged to visit Morocco and continue my pursuit of naturally occurring African amber. The plan was to fly to Rabat, rent a car, and drive to the Berber areas in the Atlas Mountains. All of the stories I had heard connected African amber to the Berbers, an indigenous Caucasian people who had lived in northern Africa since antiquity. The area had been colonized by the Romans in A.D. 50, invaded by the Arabs in A.D. 700, and attacked by the Bedouins in A.D. 1200. The Berbers themselves appeared to be peace-loving peoples that cultivated the lowlands and tended their flocks in the Atlas Mountains. They had preserved their own language, beliefs, and traditions, and excelled in pottery making and weaving.

The plane flew from Marseilles directly to Rabat, over small mountain ranges and forests of cork oak. I could see the white cattle egrets gliding back and forth beneath the small aircraft. After landing, I immediately rented a car and drove to the Rabat central marketplace. It took a while to adjust to the African scene again—the multitude of different smells and colors, women with veils, beggars—but the climate was less humid than West Africa.

It didn't take long to find the "amber." There were beads of all shapes and sizes, some as big as baseballs. I had never seen such an assortment. The strings of yellow beads were impressive against the olive-skinned women with their penetrating dark eyes. The prices were high, as much as 15 dirham for a few fake amber beads, as much as 120 dirham for a broken necklace with three plastic imitation amber beads; and at first the merchants didn't want to bargain, because they expected Westerners to pay the

A Moroccan woman with a necklace made with a mixture of plastic and Baltic amber beads.

asked price. I began to think that either the amber was highly valued or a great many tourists were looking for it. Almost all I bought or saw was plastic. But I must say that the Moroccans showed imagination in making imitation amber: not only were various types of plastic used, but also pine resin, glass, stone, and even treated camel bone.

I decided to leave all this commercial business behind and try to find some naturally occurring material. So I climbed into my rented Simca and left Rabat, driving toward the ancient walled city of Chellah. In the old city (Meridian), I stopped and spoke to a merchant about amber. This pleasant man admitted that most of his "amber" necklaces were plastic, even demonstrating the fact by burning the edges of two beads with his cigarette lighter. The plastic odor was decisive. He mentioned that Berbers mixed pieces of real amber with tea to control pneumonia, and rubbed sore backs with amber to heat the skin and soothe the muscles. I continued driving east toward the ancient city of Fes.

In Fes I drove first to the Medina, or central marketplace, hoping to find real amber since this was near one of the reported sources of African amber—the Atlas Mountains. I passed vendors selling the famous Fes plates and found a merchant who sold amber by the gram, six dirham per gram, and swore that this was the "real" stuff. I sat down in a small café and ordered a bowl of saffron-flavored soup and a plate of ragout, which I ate like the natives, with bread and fingers. As I walked through the market, watching adults and children sweating in small, dark rooms chipped out of concrete walls and completing tasks that machines, if not animals, normally perform, I felt transported through time. These laborers knew few pleasures, and that may be why marijuana (called "keef" there) was so commonly used—it dulled their stomach pains.

A lucky worker could find employment with one of the local tanneries and end up skinning cows and sheep, and then standing in huge vats, dipping the hides in red and white dyes. But he would have to contend with the odor of rotting skins, which was almost overpowering. Or he could send his eight- or ten-year-old son over to the metal shops, where the boy would spend ten hours a day bending over a workbench and hammering designs in metal plates. In the evening, when he arrived home, the stains of black metal dust (from polishing silver plates for the tourists) would render him nearly unrecognizable. When the boss wasn't looking, these poor children would quickly turn and beg for a cigarette or a dirham.

Amber items were definitely more popular here than in West Africa. Many stalls carried artificial Berber amber. I even discovered some small bottles of perfume made from mixing together cedar sap with bits of crumbled real amber.

Continuing south to El-Hujib and then Azrou, I found that the Berber influence was very strong. Many craftsmen were coming down from the mountains with beautiful handmade rugs for sale. Few Berbers did the actual selling; they usually turned the merchandise over to Arabs, who were

Berber children in Morocco.

much more skillful in business matters. The amber dealers here performed an assortment of acts to "prove" that their beads were real amber. One trick was to pick up a bead, rub it against your shirt, and then pick up a small piece of paper. Since rubbing it produced static electricity, this "proved" that it was amber—but I already knew that many plastics had a similar property. Another trick was to smear a little pine sap on the back of the bead and then heat it with a match. The pine odor is conspicuous. That the plastic piece doesn't burn with this treatment is not important, since most tourists don't know that real amber burns.

Reaching the first range of the Atlas Mountains, I was surprised to find snow in June. The rocky dirt road continued through pine and cedar forests, and then dipped down rapidly to marshy areas with giant reeds and palmettos. As I drove south, the landscape changed and the mountainous terrain grew rockier and drier. Small springs flowed out of the mountain passes and collected to form a fast-flowing river that eventually served to irrigate lower-lying fields of broad beans and barley. From time to time, a small oasis would appear with a dense but very limited growth of willow, tamarisk, and oleander.

This was the area of the Casbahs—strikingly impressive, castlelike complexes with square towers and high walls, which surround a number of small homes and gardens. The construction materials consisted of rocks held together with mud, sometimes mixed with reeds. Recently, a layer of protective plaster had been added to the walls. The Casbahs were originally fortified complexes built for defense, and most retained their original character.

The journey continued the following day to Erfoud, where the paved road ends—the land farther south belongs to the Sahara and the nomads. The next village south was the center for camel sales, and I watched from my car as a Berber boy guided his four camels in that direction.

Traders in the Erfoud marketplace will barter anything. One whipped a pen and spoon from my shirt pocket and wanted to trade a wooden carving for both. Not far off were the typical Berber tent homes with children playing outside. These goat-skin dwellings are made so they can be quickly dismantled, packed on the backs of camels and mules, and taken, along with the flocks of sheep, to greener pastures.

I continued to the village of Boudin, another typical walled city with an open marketplace just off the main highway. I went from stall to stall asking for amber but found only plastic. Suddenly, I felt a hand on my arm and, turning, looked directly into the face of a well-dressed man in his early forties, short and solid, with neatly cropped dark hair and sparkling brown eyes.

"You are looking for amber?" he asked. "I can get a very good selection for you."

"Can you get some of the naturally occurring Berber amber?" I replied.

"Oh yes, I am sure," he answered in an authoritative voice. "Come to my home and I will call my friend to come and bring you a very nice selection."

Although I felt wary, I feared that this might be my only chance to secure some real African amber. So I agreed and accompanied the man to his house, which was inside the Casbah. As we walked through the ten-foot-high solid stone fortification surrounding his house and animal enclosure, I noticed that he locked the main and only door to the road. I relaxed when he introduced me to his wife and two young boys; the more people around, the safer I felt. The ground floor comprised three rooms: one for the goats, another for dried alfalfa, and a third for his own fruit and grains. We proceeded up the flight of wood stairs to his living quarters.

The room was decorated with Arab designs and the pictures of various leaders, the center of one wall bearing the likeness of Allah. The man spoke to his Berber wife, and she quickly left the room. "We will first have our traditional mint tea ceremony," he exclaimed as he put on a robe, spread out a prayer rug, and bowed to Allah. After several minutes, his wife shoved a tray with hot water through the door and left. He sat cross-legged in front of the tea set, which he filled with boiling water. He then placed a handful of spearmint leaves, some lumps of sugar, and four spoonfuls of regular tea in the pot, and left it to steep for a minute over a candle flame. He poured it by raising the pot above his head and directing the narrow stream of greenish tan liquid straight into the small cups below. Just before the cups were filled, he dropped his hand and in a graceful curve tilted the pot to cut off the stream at the last moment. It was quite a feat.

I was growing anxious. Here we were drinking mint tea, which was fine in itself, but I wanted to see the amber and continue on my trip. My host, however, had other plans. "The amber merchant has been detained but will be here shortly, after dinner," he explained. "My wife will now prepare a couscous dinner for you. But we must have wine, so could you purchase some wine for dinner?"

What had promised to be a simple examination of amber was evolving into a much more complicated matter. Biding my time, I agreed to buy some wine, and we then walked together to another part of the Casbah. I would have left then, but he had hidden my backpack out of sight somewhere and it contained my camera and travel papers.

We proceeded down a narrow dark alley and approached a small hut. Beggars lay in the gutters. He knocked several times on a battered wooden door. The top portion of the door opened and a short, fat old woman appeared. I noticed that she had no veil. She mumbled a few words, my companion answered, and two large bottles of wine were slammed down on the upper door ledge. My friend motioned for me to give the woman the money and take the bottles. I knew that drinking was forbidden under the Muslim religion, but apparently if someone else bought it and carried the bottle, it wasn't considered that bad.

It was pitch dark outside when we arrived back at the house. Nervous, I told my host that I wanted to leave now, even though I hadn't seen the amber. He objected strenuously, saying that his friend would be here any minute, that his wife had cooked the couscous, and that we should at least eat first. Then he said not to worry about my car, because it was being guarded by his friends; this worried me even more.

The dinner would have been very tasty if I'd had the peace of mind to enjoy it, but I grew increasingly concerned as my host consumed a bottle of wine and became more and more tipsy.

Finally, the dinner was over and I explained as politely as I could that I had to leave now. He pointed a shaky finger to a small room off the one we were in and said, "Well, your backpack is in there." I got up and entered the room. It was piled high everywhere with half-made rugs or blankets. Rummaging through the textiles, I located my backpack, but before I could leave, the door swung closed and the room became suddenly dark. I made my way toward the door, but before I could reach it the key turned in the lock. Here I was, locked in a room in a village near the Sahara—all because of amber that had never appeared. I tried to think how to get out. Yelling through the thick closed door produced only a slurred response—sleep here, and maybe he would let me out tomorrow. There was one window in the room, about one by two feet, that maybe I could have squeezed through, but three solid metal bars embedded in the hardened mud sills prevented me from even trying.

I decided to sit down, try to remain calm, and think. Unfortunately, I couldn't think of anything useful. I managed at best a fitful sleep that night, periodically waking up only to discover that this frightening situation wasn't a dream. Locked up like one of my host's goats, I listened to the cries of the jackals on the edge of the desert. But everything has to end, including that night. The dawn brought with it calls from the tops of the mosques, summoning the faithful to prayer.

I had no new ideas for escaping, but the morning light brought new determination. I pounded on the door. No response. I continued to pound. Finally, the wife called: "He is not home yet. You must stay there."

"I have to go to the bathroom, that's all," I shouted, as my heart pounded.

"You have to wait."

"I can't, I have to go now."

There was silence, then the key clicked and the door opened a crack. "All right, just go to the bathroom," she ordered. In an instant I was out of the room, through the dining room, and running down the stairs.

"No, no, come back!" she screamed as she ran behind me.

My hand pushed on the gate door. It was locked! I looked around— ten-foot-high smooth stone walls. How was I going to get over those? Then I saw my chance, a small, three-foot-high fence around the outside goat pen and abutting the wall. I was up on it in an instant and had my arms over the wall. Someone grabbed my pant leg and pulled. I looked down into the dark fearful eyes of the Berber woman. "Don't go, he will kill me," she pleaded.

For an instant I hesitated, but then the thought "If he would kill her, he would probably not hesitate to do the same to me" entered my mind. I yanked my foot out of her grip, climbed up on the wall, and then hung down by my arms on the other side before dropping the final few feet to the dirt path below. As I landed, I heard an angry shout behind me and turned to see my captor approaching, waving a short knife above his head. Fortunately for me, he must have consumed the second bottle—he was staggering and looked like he had a good hangover. Too bad I didn't buy him a third bottle, I thought, as I began to run toward the market where the Simca was parked. I hoped the car hadn't been stripped. I ran around the corner and there it was, my car, still in one piece but literally covered with Arabs, some sleeping, others sitting on the hood and fenders. Shouting at them to get off the car, I began kicking away the huge rocks they had placed around the tires. I wrenched open the door, threw in my backpack, and tried to calm my shaking hand long enough to shove the key into the ignition switch. My drunken "friend" was lurching down the road toward the car, still waving his knife. I couldn't start the darn thing. Had they done something to the motor? Finally the engine caught, and just as he lunged toward the door, I pulled out of the marketplace and turned down the highway. The rearview mirror framed him standing in the street, his mouth moving and his hand still clutching the knife. I hope he didn't beat his Berber wife, since I probably owe her my life, but I will never return to find out.

A careful examination of the "African amber" I acquired on that trip showed that the great majority was imitation, mostly composed of various types of plastic. A small portion was real amber. When two of these pieces were analyzed using a method that examined chemical compounds called pyrolysis mass spectrometry, one turned out to be Baltic in origin, but the other was clearly different. Whether it was an artificial product or a piece of the elusive amber from the Atlas Mountains remains a mystery. Aside from a recent report of supposedly true amber known as Amekit from southeastern Nigeria, all other known natural "amber" in Africa is really copal (younger and softer than real amber). The most famous copal is Congo copal from *Copaifera* trees in Central Africa and Zanzibar copal from *Hymenaea* trees in East Africa. The mysterious Berber amber from the Atlas Mountains has not yet been found.

8   Bringing Them Back Alive

Back in California after a year's absence, I was ready to turn my attention to some unanswered questions that had arisen from our fossil nematode study. Could bacterial spores survive for millions of years in amber? We were attempting to answer this question when, in October 1980, we sterilized the surfaces of Baltic and Mexican amber samples, crushed the amber, and placed the powder on culture media. Our laboratory was ideally suited for conducting such experiments because we maintained an insect disease diagnostic service at that time. This service involved isolating various microorganisms (bacteria, fungi, protozoa, viruses, and rickettsia) from diseased insects that were submitted by research workers throughout the world. So we had culture media for the various pathogens, and we knew what types of organisms to expect in the different types of insects. We knew what precautions we had to take to protect ourselves from exposure to unknown pathogens and how important it was to maintain sterile conditions. We pondered for some time the best way to surface sterilize a piece of amber. One problem was that many fine fractures in amber can extend from the surface into the matrix. These fractures could contain recent, as opposed to ancient, bacteria. We chose pieces of polished amber that had no visible fractures and first placed them in a wetting agent, then in a five percent bleach solution for two hours in a vacuum oven. Previous experience told us that this treatment would destroy *Bacillus* spores. The pieces were then carefully removed and placed in 70 percent ethyl alcohol for sixty hours under vacuum. They were then rinsed in sterile water. The water rinse was cultured on the same range of media used for the amber specimens.

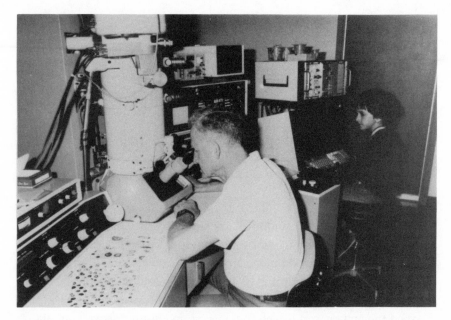

Glen Epling at the electron microscope examining "ancient" bacteria isolated from Mexican amber. Nicky Tkach is in the background.

In the original experiment, we crushed two pieces of Baltic amber and ten pieces of Mexican amber. *Bacillus* and *Micrococcus* colonies appeared after several weeks on brain heart infusion agar plates, or culture plates, set from two of the Mexican samples. None of the other samples, including the rinse controls, showed any growth. We peered down at these small colonies, wondering if they were truly ancient bacteria. There was really no way to determine whether the *Bacillus* or *Micrococcus* originated from the amber matrix or were contaminants from our laboratory. We decided to discuss this matter with other scientists.

Several years passed before we again turned our attention to these bacteria. In the meantime, they were subcultured and stored under refrigeration. When I visited Montana in March 1983, I took with me cultures of the bacteria we had originally isolated from Baltic and Mexican amber in 1980. They were given to Dr. Glen Epling, who, together with Andy Blixt, later supplied transmission and scanning electron micrographs of the bacteria. Another sample was sent to the Centers for Disease Control in Atlanta. After

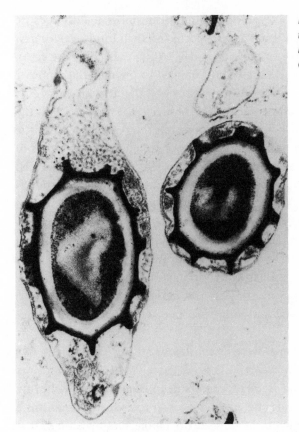

Electron micrograph of bacteria putatively isolated from Mexican amber.
(Photograph by Glen Epling)

some months the reply from Atlanta read, "We were unable to identify any of the paleobacteria to species," and suggested that we conduct further deoxyribonucleic acid (DNA) studies.

We continued to search for spores in amber over the next several years, always ending up with several species of bacteria and fungi on our culture plates. Were they all contaminants? We decided that a better method would be to try to isolate microbes from specific insects in amber that corresponded with present-day species so we would know what to expect on the culture plates.

For our experimental insect, we chose the common species of stingless bee in 25- to 40-million-year-old amber from the Dominican Republic.

These extinct bees are fairly common in Dominican amber, mainly because they collected the resin (as droplets transported on their hind legs) for nest construction, as their descendants do today. While collecting the resin, some of the bees became entangled and trapped in the material. Another interesting aspect of the biology of stingless bees is that they carry specific *Bacillus* bacteria in their alimentary tract. These bacteria are placed on the pollen food reserves in the nest and appear to preserve and prevent the pollen from spoiling. So in November 1984 we obtained twenty pieces of Dominican amber, each containing a stingless bee. This time we decided to use fumigation to surface sterilize the amber pieces, feeling that the gas would be able to enter small cracks extending from the surface into the matrix of the amber. After determining that ethylene dibromide fumigation would destroy bacterial spores artificially placed on the surface of pieces of polished amber, we subjected our experimental pieces to thirty hours of ethylene dibromide gas in the fumigation chamber at the Lowie Museum of Anthropology on campus. The museum used fumigation to destroy wood-boring insects and fungi associated with carvings and other wood artifacts.

Of the forty cultures that were established from this experiment, five were positive for bacteria and two of these turned out to be bacterial species of the genus *Bacillus*. Had we been successful in isolating and culturing ancient bacteria? We turned the cultures over to a graduate student in a biochemistry laboratory who was working on a molecular revision of that particular bacterial genus. He identified both as *Bacillus subtilis.* On the one hand, this was exciting because this species of bacterium has been recovered from the alimentary canal of modern-day stingless bees. On the other hand, this bacterium is also a common soil inhabitant and therefore could have come from anywhere. We didn't know how to determine if these bacteria were truly ancient.

So we never could prove that what we isolated from amber over the years were real ancient bacterial species. This line of work is being continued by Raul Cano at California Polytechnic State University in San Luis Obispo. He has made progress by showing the presence of *Bacillus* DNA in the abdominal cavities of stingless bees in Dominican amber. We anxiously await further results of his investigations.

9   Million-Year-Old Cells

Few things rival the beauty of fossils in amber. When examining insects entombed in this gem of the sun, you can hardly be unmoved by the wonder of seeing marvelously preserved invertebrates frozen forever in their everyday tasks—worker ants carrying food, bees with outstretched wings carrying pollen, flies mating. But it is even more miraculous, as we discussed earlier, that under the dissecting microscope one can easily see in some specimens intact tissues, remnants of their internal organs. This alone makes amber-embedded fossils unique, because few other fossils have been found in which soft tissues have been so well preserved. It was therefore only a matter of time before the right circumstances presented themselves—a well-preserved fossil with tissue, an experienced microscopist with an electron microscope, people with a love of amber and the willingness to look—and the idea that the fine structure of these internal tissues might also be preserved would be born and ultimately pursued, especially when our previous studies had so strongly suggested it to be true.

That someone as involved in the study of amber and amber fossils as I am would share their magic with his or her family is only natural, and amber really is a family affair in our case. One day in 1980, Roberta and I were taking turns peering down the oculars of a dissecting scope at a particularly well preserved set of Baltic amber fossils. The pile glittered in the sunlight, really fossil gold. Typically, when we are sharing some very special amber fossils, we utter many of the same exclamations of delight and wonderment that you would hear at some awe-inspiring event such as an especially vivid fireworks show. When that soon-to-be-famous female fly appeared under the spotlight of the dissecting scope, we cried, "Oh my

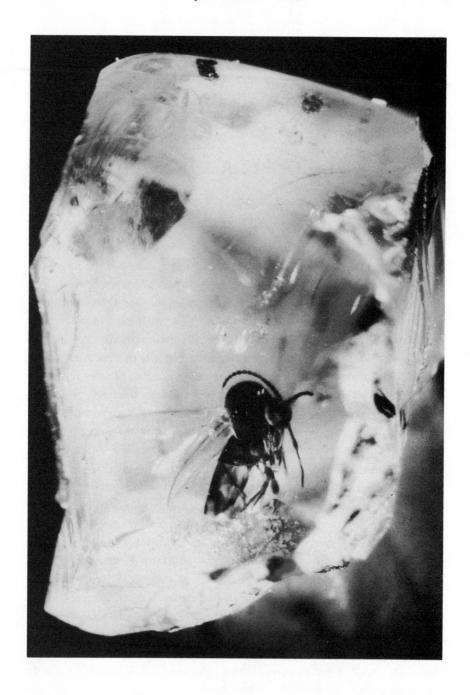

God, look at this!" There was so much tissue, so well preserved. That fly looked like a freshly embedded specimen ready for a microscopist's knife. When we looked up at each other, the same thought was written across our faces—surely this fly's cell structure must be intact too! From then on, that mycetophilid fly became the center of our research project.

The day finally arrived when Roberta could begin work on the fly. We hoped that all the bugs—excuse the pun—had been worked out of the system in the mid-1970s, when we had worked on the fossil nematodes. I had already photographed the intact specimen, and a small piece of the amber had been chipped off and sent to Dr. Curt Beck for infrared spectroscopy analysis and confirmation that our fly was indeed in Baltic amber. We had decided to break the fly into two pieces and save one half for future studies. The piece for safekeeping was stored in an airtight container in the dark. The specimen had broken through the abdomen, and Roberta started to work on her piece. The tissue was obvious as a dark strip adjacent to the cuticle. The center of the fly, however, was an open cavity, and this proved to be the first hurdle in obtaining ultrafine sections. After a frustrating day attempting to obtain any usable sections, we decided that we would have to fill the cavity with embedding plastic before we could go any further— just another delay. It was several days later that Roberta was finally able to look at our tissue in the electron microscope. When I returned to my office after a trip to the library, a note was hanging on the door: "Success!" I went immediately to the microscope room and breathlessly looked down at the lighted screen, straight onto tissue that was 40 million years old. Tissue with nuclei and organelles—mitochondria, lipid droplets and ribosomes— was there before my eyes, as well as entire muscle bands with easily identifiable components such as fibrils and mitochondria. Tracheoles, the breathing apparatus of insects, had intact linings, recognizable tubercles, and even possibly remnants of the plasma membrane.

There is something almost spiritual in a discovery such as this. It certainly makes one feel humble to look on intact cells that have been around for 40 million years. Long before human beings were even considered in

◄
The famous Baltic amber fly.

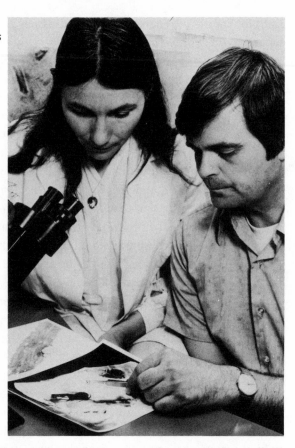

*The authors examining
photographs of the famous
Baltic amber fly.*
(Photograph by Pat Craig)

the evolutionary time scale, cells existed with the same structures and organelles that they have today. How insignificant and mundane we really are.

The structure of the cells seen in our amber has a different appearance than that of cells standardly prepared for electron microscopy today. This is because scientists strive to obtain what they regard as the best fixation with cell structures having clarity and definition. Some say that electron microscopy is a study of artifacts—that tissues are frozen in time with harsh chemical fixatives, and cells are viewed when caught in a millisecond

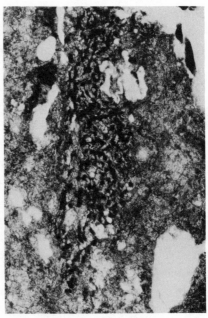

Electron micrograph of tissue from the 40-million-year-old Baltic amber fly.

Electron micrograph of 40-million-year-old nuclei from the Baltic amber fly.

of life, with their whole life cycle extrapolated from these images. There are obviously many parallels between studying cells with the electron microscope and studying insects entombed in amber. But in the case of amber, nature used its own methods of fixation and dehydration. We feel that the fossil fly best resembles modern-day tissues that have been processed by inert dehydration using ethylene glycol. Sugars, terpenes, and other compounds in the tree sap may have combined with water in the cells to dehydrate and preserve the tissue of entrapped insects—almost a mummification process. In the past, humankind was aware of the preservative qualities of tree resins. The Egyptians used resins to embalm their nobility and wealthy citizens. Resins have antibiotic qualities that destroy fungi and bacteria and retard decaying, plus they have components that preserve tissue. Myrrh, a common embalming agent, is a mixture of the resin, gum, and essential oils from the *Commiphora* plant. This was poured into the cranial, chest, abdominal, and pelvic cavities of ancient Egyptians, and the bandages that wrapped mummies were soaked in it. Resins have also been used topically on wounds as an antiseptic and in wines to prevent spoilage. Greek Retsina wines still use resin for flavoring.

Our study of the fossil fly continued over many weeks and followed the routine cycle of ups and downs associated with electron microscope research. Although some areas of the fly were well preserved—especially a strip of hypodermal tissue immediately adjacent to the undersurface of the fly's cuticle—other tissues, further into the body cavity, lacked substructure and were composed of ghostly cell outlines. Muscle cells were the best preserved, and that has been observed by researchers in other studies of ancient tissues, such as mummies or woolly mammoths.

Preliminary results of our studies were first sent off in July 1981 to the journal *IRCS Medical Science,* where they were published later that year. More conclusive results after months more of study were published in *Science* in March 1982. This paper represented a pivotal point in our research plans. If tissues could be discovered so well preserved in amber-embedded insects 40 million years old, what else could be found? If the nuclei contained chromatin, a darkly staining structural component known to include genetic material, could DNA (deoxyribonucleic acid) actually be there too, just waiting to be discovered? If there were ribosomal structures, why then not ribonucleic acid (RNA)? We wondered if insects in older amber also had well-preserved tissue remains. On March 5, 1982, we wrote and asked Frank Carpenter at Harvard if he could send us an insect fossil in Canadian amber. On May 12 he sent us a braconid wasp from 70- to 80-million-year-old Cedar Lake amber. Roberta treated the specimen in exactly the same manner as she had treated the Baltic amber fly. The Canadian amber also sectioned with the glass knives and under the electron microscope, the sections showed well-defined tissue in the wasp's abdomen. Although the few sections we examined did not appear to pass through any nuclei, we did observe distinct sections of trachea and surrounding laminated membranous structures in partially vacuolated cytoplasm. We now had evidence that amber from different ages and plant sources could preserve insect tissues, cells, and cell organelles.

Due to our work on fossil bacteria, we had unprovable evidence, but evidence nonetheless, that even life forms might lie dormant entombed in amber. What about nucleic acids? As we were asking ourselves these questions, unknown forces were at work in other places, fitting together the pieces of a complex pattern of techniques, knowledge, and expertise that would ultimately enable us to find answers to these questions.

10   Pursuing Ancient DNA

When you can't answer a scientific question, you consult an expert who can. This principle has been the foundation for all of my research in amber. In the area of molecular evolution and paleontology, Dr. Allan C. Wilson was the leading expert in the early 1980s, and the logical one to turn to for an answer to our questions about DNA in amber. I was fortunate that he was a colleague at Berkeley and therefore was located only a couple of buildings away. But even as I was preparing to seek him out, he turned up at my office one day in spring 1982 with a reprint of our *Science* paper on the ultrastructure of the amber-embedded fly in his hand. He was very enthusiastic about our findings and expressed an interest in a collaborative project. Dr. Wilson, a faculty member in the Biochemistry Department, was a pioneer in the study of ancient DNA, and by the 1980s his laboratory was looking at the DNA of several recently extinct animals. Wilson and Dr. Russell Higuchi were able to extract and clone pieces of DNA from the 140-year-old pelt of a quagga. The quagga, *Equus quagga quagga,* once roamed the plains of southern Africa but was driven to extinction by excessive hunting by white settlers in 1878. The relationship of the quagga to the horse and the zebra had been hotly contested by scientists for some time. By extracting fragments of DNA from the quagga and comparing it with those same areas of DNA from the horse and zebra, Wilson's laboratory determined that the quagga was more closely related to the Burchell's zebra. Other ongoing projects in that laboratory included looking at DNA from a hundred-year-old skin of a bison and from a Holocene bone of an extinct moa. The moa, a tall, flightless bird once found in New Zealand, also owes its extinction to humankind. The Maoris arrived

The late Allan Wilson, the father of research in ancient DNA.
(Photograph courtesy of Leona G. Wilson)

in New Zealand in approximately A.D. 750. Archaeological sites with moa remains indicate that the moa was present as late as 1350, but it was gone from New Zealand by 1624—obviously hunted to extinction—when the first Europeans arrived. The moa was similar to the present-day emu of Australia and the ostrich of Africa. DNA obtained from the bones of the moa would provide information about its evolutionary history and relatedness to other birds.

Wilson's laboratory was also looking at a 40,000-year-old mammoth, found frozen in Siberia, that contained fragments of DNA preserved in muscle tissue. Roberta had the opportunity to look at some of the electron microscope studies done on the mammoth, through her friend Alice

Taylor, the electron microscopist for the Biochemistry Department at that time. We noted that, compared with the insects in amber, very little structure was recognizable in the mammoth except for the collagen fibers preserved in the muscle.

These and other projects had put Allan Wilson in the forefront of paleo-DNA research, and we had in fact already made preliminary contact with people in his laboratory about the possibility of finding DNA in amber-embedded insects in August 1981, after we had our initial results on the electron microscopy of the fly. At that time I met with some of the people in the laboratory, including Russell Higuchi. The idea of extracting DNA from amber-embedded insects was perhaps a little before its time then, possibly because the technology wasn't yet in place. At any rate, nothing had come out of the first meeting. (At about the same time that we were beginning to investigate fossil DNA in amber, Dr. John Tkach of Bozeman, Montana was corresponding with people with similar thoughts. In February 1981 he had written to Wilson suggesting that fossils in amber would be a possible source of ancient DNA. We first contacted Tkach in December 1982—but that story will come later.)

Our second contact with Dr. Wilson himself was more fruitful, and we finally coordinated a project to look for DNA in amber. We started to work with Russ Higuchi in early 1983. We first needed to locate amber specimens that we thought might have significant amounts of tissue present, and it wasn't until the first week of April that Roberta prepared the initial fossil amber samples for DNA extraction. Her handling of the amber in the earlier electron microscope studies had given her some invaluable experience with cutting and otherwise processing the pieces. Previous work using sterile techniques was also valuable. Russ Higuchi had stressed the importance of not contaminating the amber pieces with foreign (not fossil) DNA. For example, the cell of a *Drosophila* fly may have 2 picograms (1 picogram = 1 trillionth of a gram) of DNA, but a cell from human skin may have 5 picograms. Therefore, contamination of the amber sample by sloughed-off human cells would constitute a serious problem. We would also have to contend with ubiquitous bacteria and fungi. The trick was to be as clean as possible and to avoid any steps that could lead to picking up contaminants.

Preparation for this phase—getting the tissue out of the amber and into the extraction solution—took several days. It is amazing how many things are handled by humans and therefore could have traces of human DNA. The dissections of the amber were first done by roughly cutting the amber as closely as possible to the insect with a hacksaw similar to those used to trim specimens in the electron microscope study. These blades had to be washed, flamed, packaged in paper that hadn't been handled by anyone, and sterilized in an autoclave. The area where the samples were to be prepared had to be cleaned and readied. The final stage of tissue extraction—the trimming away of the amber with razor blades to produce a small hole in the insect body cavity—had to be done under a dissecting microscope. The microscope also needed to be decontaminated and cleaned—first by washing with an antimicrobial solution, and then by setting the instrument in a box with ultraviolet light. Dissecting tools, razor blades, needles, gloves, lab coats, trays, vials, picks, forceps—everything had to be washed, handled with clean gloves, packaged, and sterilized.

In between these preparations, Roberta practiced cutting amber. She soon discovered, after receiving a nasty cut from the razor, that the small pieces of amber would need to be secured by a vise for the final close trimming. We located a suitable vise, finding that the same vise-type chuck used in the ultramicrotome (the machine used to produce ultrathin sections for electron microscopy) to hold the specimen blocks was also the one best to use in this experiment.

Myriad issues come up when you do a procedure for the first time. Occasionally, you have blind luck—everything sails right along and the experiment is successful. Usually, however, it's an uphill fight, with everything that can go wrong, going wrong. Sometimes the problems are major, such as finding a vise lying around your lab somewhere that will hold a small piece of amber. Other times small delays hold you up, such as having the edge of your razor blades snag and tear holes in the gloves you are using.

Seven pieces of amber that appeared to have tissue were selected and trimmed close to the insect. The razor blades were flamed, the amber was surface sterilized, and the tedious task of chipping away at the amber began. The first sample, a moth, contained tissue that appeared white. When the razor entered the body cavity, it was seen that resin had infiltrated it. Although the interior was chipped out, no DNA was subsequently recovered from this sample. The second sample was a spider. The body cavity here was full of liquid when opened, and in later work no DNA was recovered. The other five samples were filled with dusty tissue remains, or black tissue fragments adhering to the cuticle. In two samples, a moth and a fly, Russ Higuchi detected putative fossil DNA. He did this by using a "template assay," a technique used to produce multiple copies of fossil DNA. The fossil DNA served as the template and was used as a guide upon which the base components of DNA—adenine, thymine, cytosine and quanine (supplied in a radioactively-labeled solution of nucleoside triphosphates)—were laid down. The ends of the fossil DNA templates were prepared by the addition of primers (exogenous oligonucleotides) which supplied the information needed to initially engage and move the replication process. The actual copying procedure was driven by purified DNA polymerase I (from *Escherichia coli*) which bound the base sequences and tied the new strands of DNA together. At that time, the template assay could detect as little as 10 picograms of DNA. This assay was a predecessor, so to speak, of the now-famous polymerase chain reaction (PCR, patent licensed to Perkin-Elmer Cetus), which has revolutionized DNA research.

The polymerase chain reaction was introduced commercially in 1987 for amplifying nucleic acids. It uses high temperature to denature and disassociate double-stranded DNA into single-stranded DNA. These single strands are in a buffer solution with an excess of specific primers that anneal to sites flanking the region of DNA the researcher chooses to amplify. In addition, the solution contains deoxynucleoside triphosphates, the building blocks of DNA and DNA polymerase. A complex is formed between the DNA template and the ends of the annealed primer, and the sample DNA is copied between the primers in a polymerase-mediated extension. For each original piece of fossil DNA, there are now two double-stranded pieces of DNA, each with an original piece of DNA and a new piece of amplified DNA. The process is successively repeated until adequate copies of that specific fragment of the sample DNA are obtained for sequencing and cloning. A thermostable DNA polymerase, *Taq* DNA polymerase isolated from the hot springs bacterium *Thermus aquaticus,* is now used in PCR. This improved thermostable polymerase stays active during repeated exposures to the high temperatures used for denaturation and does not need to be added each time as was necessary for the DNA polymerase I—a tedious and expensive process.

PCR is particularly useful where only microquantities of DNA are present and need to be amplified before standard research techniques can be applied. This innovative technique can be applied to the study of fossil DNA where only small amounts of DNA are preserved, but the technique is rapidly expanding into many fields such as forensics and medical diagnostics.

Kary B. Mullis won the Nobel Prize in 1993 for the invention of PCR. Dr. Mullis related the circumstances that led to the discovery of the PCR in the April 1990 issue of *Scientific American.* He was hired in 1979 by the Cetus Corporation, then located in Emeryville, California, to synthesize oligonucleotide probes. The subsequent automation of the synthesis process left him with some free time to develop his own ideas, and so he started "puttering around with oligonuceotides." The actual concept of PCR occurred to him as he was driving on a Friday night in April 1983 along Highway 101, headed toward Mendocino and the Anderson Valley. His companion was sleeping, so, as he drove, his thoughts turned to his ideas for a DNA-sequencing experiment. While the miles slipped away, the design of his experiment continued to change as he considered the feasibility of first one method and then another. Near the end of his journey, he realized that the technique he was formulating in his mind would exponentially amplify fragments of DNA, and he became quite excited. A little farther along the road, he realized that he had devised a way to fix the length of the DNA strands he would be amplifying by designing specific primers. By then his enthusiasm was monumental, and rightly so. It is fortunate that his journey wasn't shorter and that his companion slept during the trip; otherwise this idea may never have been completely developed. As Dr. Mullis points out, given the conceptual simplicity of PCR, "the fact that it lay unrecognized for more than fifteen years after all the elements for its implementation were available strikes many observers as uncanny."

Although seemingly positive results of the preliminary DNA studies from amber-embedded insects were mentioned in an abstract published by Russell Higuchi and Allan Wilson in 1984, the project was not actively continued in their laboratory at that time due to a lack of funds. In autumn 1984, Dr. Wilson had submitted a grant application to the National Science Foundation (NSF). Part of his research plans included looking at the DNA in amber insects, specifically, *drosophilid* flies. It's too bad that his NSF

Svante Pääbo, the first to obtain DNA from an Egyptian mummy, at the 1993 ancient DNA conference in Washington, D.C.

proposal was turned down, since Wilson's lab was to be among the first to use the PCR technique; we might have progressed faster than we ultimately did toward our goal. Nevertheless, we maintained contact between our labs, and exchanged and discussed ideas and techniques over the next few years. (Interestingly, Dr. Rob DeSalle was beginning postdoctoral work in the Wilson laboratory at that time. He would head the team that later extracted DNA from a fossil termite in amber.)

Several years later, in autumn 1989, a second attempt to isolate DNA from amber-embedded insects was made by Dr. Svante Pääbo, then a postdoctorate fellow in Allan Wilson's lab, known for his research on Egyptian mummies. In this attempt, bees in Dominican amber were examined. The amber was shaped as closely to the insect as possible and then thoroughly washed and rinsed. Next it was transferred to a mild acidic solution for several minutes to remove any adhering cells, neutralized, and placed in tubes. The amber was then crushed in the tubes, and extraction fluid with all the necessary components was added. Unfortunately, no DNA was detected in these experiments. In retrospect, perhaps the crushing of the amber and the fossil insect together somehow bound the DNA or interfered with the process of extraction. Of course, that no DNA was present in those particular specimens is also possible. Obviously, not every sample contains fossil DNA.

At any rate, we were disappointed with these negative results after the initial positive ones. Dr. Pääbo was leaving Berkeley, and no one else seemed to be interested in looking for DNA in amber insects. Allan Wilson, a staunch supporter of the project, was ill with cancer and involved with other projects. (He passed away in 1992 at the age of fifty-six.) No funding was available for our amber research, and this line of experimentation seemed to be at a dead end. But we felt there was still ample reason to believe that DNA was present in amber-embedded insects. First of all, we knew from the electron microscope studies that the nuclei and mitochondria, both DNA-containing organelles, appeared to be preserved, perhaps fixed. We now know that DNA in specimens prepared with formaldehyde, acetone, and other fixatives can indeed be extracted and amplified by PCR. These fixatives are thought to act by cross-linking proteins (formaldehyde) or by precipating them (acetone), and similar mechanisms may occur when an insect is preserved in amber. Aldehydes only slightly react with nucleic acid, and it would appear that cross-linking the proteins around the nucleic acid immobilizes and protects the latter. On the other hand, precipitating fixatives such as alcohols and acetone are often chosen for nucleic acid retention. We haven't any idea what type of mechanism might have been at work in the resin. The tissues may have been rapidly dehydrated by resin components. Extremely dry conditions, as well as the lack of oxygen, are all conducive to DNA survival. The native, double-stranded DNA molecule is believed to be more thermodynamically stable than most proteins at or near neutral pH, resisting denaturation and hydrolysis. Also, as the amber polymerizes, acids in the resin become immobilized and do not invade the body cavity of the fossil—another reason to suspect that DNA might still be present. And finally, resins are known to have antimicrobial properties that would prevent bacterial degradation of embedded tissue.

The project was put on a back burner. Once again, we awaited the right time, the right person, and the right expertise.

II   Amber South of the Border

In the meantime, the lure of other fossiliferous amber deposits got the better of me. Although none of the New World amber sources reveals a history as fascinating as that associated with the Baltic deposits, Mexician amber does have an interesting past.

Documents written by the Spanish missionaries who traveled through Mexico in the sixteenth and early seventeenth centuries revealed a variety of uses of amber. Obviously, the color and sparkle of this mysterious substance promoted the idea that it had magical powers. Thus local craftsmen carved raw amber, furnished by Indian gatherers, into various shapes to protect infants from the evil eye. On the arrival of the Europeans, the shapes began to include small crosses. These, and rosaries, together with some copies of the original Mayan and Aztec motifs, are created for tourists today.

Although amber necklaces were popular, another use of amber appears to be unique to this part of the world. In pre-Columbian times, amber adorned the faces of both men and women, a use that involved the mutilation of certain facial organs. Perhaps the most common, especially among the Maya, was the practice of piercing and removing the septa separating the two nostrils and inserting in its place a flattened, circular piece of amber. An early missionary commented that the practice made their noses stick out like elephant trunks—maybe a sign of nobility, or possibly of beauty. In addition, the Lacondon Indian women had their ears perforated and amber disks inserted in the lobes, thus providing a type of permanent earplug. Not to be outdone the men devised amber lip plugs, which could

Frans Blom, historian of the Mayan civilization, who led some of the early Mexican amber expeditions.
(Photograph by J. Wyatt Durham, 1956)

be inserted into slits made through the lower lip. This later became a symbol of bravery in battle.

The source of this amber was the southernmost Mexican state of Chiapas. From there, the amber traveled south to Central America and north as far as Oaxaca, Mexico. The Indians were able to trade amber for knives, lance blades, needles, bells, rabbit skins, and bird feathers. Amber chips were burned as incense in the Mayan temples, a practice later continued in Catholic church services in that part of Mexico.

Exactly when Europeans first visited the amber mines is difficult to say, but Friar Alonso Ponce may have been the first to see amber being taken out of the ground, in 1586. The discovery of Mexican amber by the modern scientific community occurred much later. George Kunz mentioned it briefly in the late 1800s, but John Buddhue, in 1935, was one of the first to publish the fact that this amber contained fossil insects and plants.

The University of California can claim a historical connection to Mexican amber because of the past activities of various scientists from Berkeley who collected and studied amber from Chiapas. The Museum of Paleontology at Berkeley harbors the largest collection of Mexican amber in the world, most of which has been cut and trimmed into tiny pieces, each containing one or more insect fossils. How the university acquired this valuable collection can be traced back to the endeavors not only of biologists but also of an anthropologist-archaeologist by the name of Frans Blom. This Copenhagen-born scientist, as a member of an oil-surveying group, entered the southern wilderness of Mexico in 1919 and became fascinated with the centuries-old remains of the ancient Mayan civilization. He broke off from the oil-surveying group and struck out on his own, first by leading small expeditions in search of lost Mayan cities. He eventually obtained help from the Mexican government and became one of the foremost authorities on Mayan civilizations. His most amazing find was his discovery not of the famous Mayan remains at the Palenque ruins but, in 1948 in the unexplored jungle, two complete altars and the last living remnants of the Lacondon Indians (Lacondonlyace), a branch of the original Maya. These few dozen families lived in nearly Stone Age conditions. Blom took some of the youngsters back to his home in San Cristobal de las Casas and taught them Spanish. After Frans Blom passed away in 1963, his wife, Gertrude— a Swiss journalist and explorer he had met during a jungle expedition— continued his efforts to preserve these ancient peoples.

Paleontologists, however, remember Blom not for his impressive work on the ancient Mayas but for his discovery of amber and subsequent assistance given to Berkeley scientists in obtaining it.

Before he died, Blom typed up some of his notes, illustrated with maps and photographs, dealing with his search for amber in and around the city of Simojovel, Chiapas. In this historical document, Blom explains how he first found amber in a landslide near the Santa Catarina River in July 1922 while working for a group of oil geologists. The Indians in this area were carving amber pieces into charms used to ward off the evil eye (spirits). The best charms were those carved from purely transparent yellow amber—and the craftsmen took great pains to file away all enclosed insect and plant remains, which were considered impurities! Not until nearly three decades had passed did Frans Blom became an official amber hunter, after Ray Smith of the Entomology Department at Berkeley commissioned him to find fossilized amber in Simojovel. At that time gravel roads were the best you could hope for, and they never seemed to continue far. Thus Blom had to travel primarily by Jeep on rocky dirt roads that were only barely negotiable during the dry season. In his journal Blom describes passing by the city of Ixtapa, where the natives produced salt from brine springs by using their homemade evaporation process. Then, at Soyalo, someone had made a small figure of a saint, San Miguelito, and placed it in a box, stating that it had the power to cure the sick, ease all types of suffering, and predict the outcome of business matters. Thousands of people flocked to see this miracle and pay for the opportunity to ask it for favors. Some Protestant missionaries from the States were campaigning against Saint Miguelito and had aroused resentment among the villagers, making it dangerous for any foreigner to remain in the town, especially at night.

Blom continued southward and noted the colorful woolen zarapes and embroidered blouses of the Tzotzil Indians who were heading for the Sunday market in Bochil. Whole families carried produce, mostly oranges, to sell. After constantly getting stuck on the high ridges between the ruts in the dirt road, and skirting around a passenger bus with a broken axle in the middle of the road, the crew finally reached Simojovel. It may have been the "amber city" to Blom and his team, but at that time the town was best known not for amber but for its production of coffee and tobacco. Some 2,300 inhabitants lived in Simojovel then, most working in the rich fertile valleys. In March 1953 most of the town still buzzed with the news of the great landslide of October 1952, when early one morning the mountain on the south side of town had moved down the slope, carrying everything with it. Two brothers related how the palm-thatched house in which

▶
Landslide that exposed amber in Mexico.
(Photograph by J. Wyatt Durham in 1953)

they were sleeping started to slide. They leapt out of their beds and ran toward the opposite mountain, just in time to see piles of rocks and debris cover their home and cornfield. Corn, or maize, was the Indians' staple food and their lives were closely tied to their cornfields, so this was indeed a tragedy. The landslide continued during the day, cracking trees and destroying homes. Occasionally, a home would move with the slide, along with its chickens and pigs, and still remain standing, as if protected by some miracle.

The landslide exposed the limestone beds beneath the soil that contained amber. These amber layers, darkened from the remains of carbonized wood, lay between a buff-colored layer of coarse-grained sandy limestone and a blue layer of sandy limestone, with only a few feet of dark humus overhead. All of the beds contained numerous marine fossils, of which scallop shells were especially noticeable. Some larger amber pieces had oyster shells still attached to them, indicating a period when the fossilized resin was beneath the sea.

Blom collected some amber samples from these exposed beds and then returned to the village. There he met with townspeople who had collected amber containing insects in the past and were now willing to sell them. These specimens, together with the samples Blom had collected, were sent back to the University of California and sparked such interest that in 1953 a group of scientists journeyed to collect more amber and map out the stratigraphy. All of this material is now housed in the Museum of Paleontology at Berkeley.

My visit to Simojovel and the Chiapas amber mines occurred some twenty-odd years later, when I was invited by Lauren and Edward Zarate to investigate the possibility of biologically controlling the local kissing bugs. The bugs were vectors of the dreaded "chagas disease," caused by trypanosomes (small flagellate protozoa) that enter the feeding wound of the bloodsucking bugs and develop in the heart, muscles, and central nervous system of humans. The heart injury can lead to death, and no suitable medication was known at the time.

▶
J. Wyatt Durham and other University of California scientists on a 1953 expedition to locate Mexican amber.

A Lacondonian Indian woman at Na-Bolom in San Cristobal de las Casas, Mexico.

My plane from San Francisco to Tuxla Gutierrez on October 25, 1982, was crowded with hordes of Mexican laborers who wanted to board a plane home before the airlines went on strike on November 1. A driver picked me up at the airport, and as we made our way over the windy, bumpy roads, past the remains of wrecked vehicles and colorfully dressed Indians walking proudly along the roadside, it seemed that things hadn't changed all that much from when Blom had written his journal thirty years earlier.

I had the wonderful luck to stay at the Villa Na-Bolom, established by Frans and Gertrude Blom as a center for Mayan culture. On the grounds stood a central office, a museum that contained fascinating artifacts from the Mayan period, and a wonderful library containing many important historical works dealing with that part of the country, some of which were handwritten accounts of the travels of monks and friars through the area in the sixteenth and seventeenth centuries. Surrounding the impressive buildings were a number of smaller structures, all built in traditional colonial Spanish style and form, with tile roofs and profusely adorned with tropical potted plants in earthen vessels.

The next morning, as I sat down for breakfast at an outdoor table, surrounded by fragrant flowers, I glanced at a neighboring table directly into the dark, intense eyes of a Lacondon Indian woman. In fact, all of the people sitting at that table were Lacondon: they were the children that Frans Blom had taken from the jungle in the 1950s to teach them Spanish—but they had grown up and now had their own children. The Indians avoided looking at us and hid their faces when cameras were turned in their direction. The women's hair was pulled back tightly, and the men wore loose, shoulder-length hair. Seeing them thrilled me and transported me back into the past, but at the same time it saddened me because these people were some of the few last remnants of the great Mayan civilization. And who knew how long they would survive in this changing world? Through an interpreter, we learned that they referred to the kissing bugs we were studying as *Juks,* and that the *Juks'* bite was quite painful. The Lacondons knew about the amber, but recent resin taken from pines and other trees growing in the area interested them more. These Indians built bowl-like incense burners out of pottery. Each pot had a reptilian-shaped head, which represented the god Hatchak'yum, who inhabited the ruins of the ancient Mayan city Yaxchilan. Hot coals were mixed with the resin to produce an incense used during religious services while food was offered to the god.

The evenings at Na-Bolom were an education in themselves. We never knew who would be at the dinner table, but they were all adventurous types who came from all corners of the world. Many had dropped out from their ordinary jobs. A banker who collected beads as a hobby was leaving his position to study trade beads in America. An Englishman from Birmington simply wanted to travel through Mexico, and a German from Hamburg who had taught German literature now wanted to spend the rest of his life traveling. Some guests were scholars looking for undiscovered Mayan remains in the jungles. One man from Wisconsin had stopped to work as a farmer and was suddenly forced to leave the village he had chosen when the neighboring volcano erupted and covered everything with ash.

On October 29, 1982, Lauren and Edward Zarate took me over to Simojovel to visit the amber mines. The town was probably much larger now

*Mexican amber miner at the Palo Blanco
mine.*
(Photograph by J. Wyatt Durham, 1953)

Mexican amber miner in 1982.

than when Frans Blom first visited in 1953. The rows of colored stucco houses lining the roads were inhabited mainly by typical *ladinos* (Indian-Spanish descendants who have adopted Western clothes and customs).

More amber mines seemed to dot the region now. About eighty families reportedly were involved in searching for amber, each owning its own mine. A 1981 government proclamation allowed mine owners to use force to protect their property from thieves. We were strongly advised to pay an official local guide to take us to the mines so that the guards would be alerted and we wouldn't be shot on sight.

A guide from the mayor's office led us up to the mountainous area that contained the amber mines. Jungle plants grow quickly in this subtropical climate and had already taken back part of the path, so our trek was tough; everyone who travels these paths carries a machete. We passed through a small portion of tropical rain forest, with huge trees covered with epiphytes and bright red parrots flying overhead. Huge silk spiders sat waiting in their giant webs all along the path, and reportedly attack anything that they entrap, including small birds. Beautifully colored butterflies gingerly landed on a host of tropical flowers, their delicate threadlike legs tickling the moisture-laden petals. Hordes of a type of large spiny caterpiller cov-

Home of a Mexican amber worker.

ered the trunks of trees, and as soon as they reached full size, they would
be roasted and eaten by the Indians.

The path continued climbing out of the rain forest and into a more
open, but still hot and humid, area. The rocks along the path were slippery
and covered with ash and fine sand from the volcano. Farther along we
were challenged by a guard with a rifle, and our guide proved his worth.
We were walking on a small path along the ledge of a hillside that dropped
steeply more than 500 feet on one side. Finally the path grew a bit broader
as it approached the first amber mines. The number of rifles leaning against
the mountainside next to the mine entrance indicated how many miners
were inside. To enter the mines—tunnels that stretched sometimes 200

meters into the sides of the mountains—you must stoop and then crawl along on your hands and knees. The dark coal and amber vein is wedged between two lighter beds of limestone. There were small pieces of amber in front, as well as all along the tunnel. We chipped a number of pieces out of the rocks and photographed the entrances.

One miner emerged for a rest and appeared upset to see us, thinking we were thieves, but he calmed down when our guide explained that we just wanted to visit the mines and collect some amber. According to the miner, amber in the local area was used now for the same purposes as in the past—to make charms, especially for the children, for protection against the evil eye (Ojo). The miner also burns amber powder on a hot piece of wood and lets the smoke waft over his aching limbs for relief.

On the way back to the village, we encountered German, Mexican, and American businessmen waiting at the guard station for the miners to return. They bought amber directly from the miners before the miners returned home. For this reason we had a difficult time purchasing amber with insects when we went to the miners' homes. Even at that time, Mexican amber, especially those pieces with fossils, could be easily sold at high prices. Through the efforts of our guide, we did find some miners who hadn't sold all of their fossils, and we gratefully purchased the little we could find.

In 1982 some merchants were selling plastic charms as amber. This industry has grown in the past ten years, and now it is not uncommon to be offered large pieces of colored plastic containing a range of insects and other arthropods, even scorpions, inside. The insects are real, but they are not ancient.

12 | The Extinct DNA Study Group

We spent autumn 1982 examining the amber pieces collected in Mexico, reviewing the results of our scientific studies, planning future amber research projects, and collecting fossils needed for our ongoing DNA investigations. Then, unexpectedly, we received some moral and technical support from members of the Extinct DNA Study Group, established in spring 1983. This group could not have been formed without the enthusiasm, energy, and organizational skills of Dr. John (Jack) R. Tkach, who was, and still is, a medical doctor practicing in Bozeman, Montana. As he mentions in his September 1993 paper "A Brief History of the Extinct DNA Study Group," his side interest in paleontology extends back to when he was a ten-year-old boy who spent his summer mornings examining the vast animal collections, especially fossils, at the Smithsonian Institution. This attraction to fossils and, especially, dinosaurs never left him, and he became more and more interested in the extinction of the dinosaurs over the years. He even proposed a theory for their demise, based on their type of immune system. We were surprised and pleased when Jack Tkach first contacted us in December 1982. His interest in ancient DNA antedates that first letter he wrote to me regarding our *Science* paper on the ultrastructure of the fly in amber. Jack Tkach had already corresponded with quite a few people about his idea of extracting dinosaur DNA from blood cells found in the stomachs of bloodsucking insects embedded in amber. Although his idea aroused interest, I'm not sure anyone believed it possible until he wrote to me. It's amazing when you consider it, how people can arrive at the same idea (extracting DNA from amber insects) from different perspectives. As our correspondence continued into the next year, he suggested some people he thought would be helpful in our amber projects, particularly the bacterial studies. Most of those people were subsequently involved in the study group. In January 1983 he agreed to edit a newsletter detailing our ideas and goals. By January he had begun asking people to join the study group. Although many of the people contacted were generally excited by the ideas and concepts of the Extinct DNA Study Group, few were willing to jeopardize their careers or suffer the derision of colleagues, so the actual membership was limited to a courageous handful.

The first *Extinct DNA Newsletter* appeared in February 1983. The study group's purpose, as outlined in the newsletter, was to establish a forum to discuss the following: the recovery and transcription of DNA from extinct organisms, the use of recombinant DNA technology, the study of protein evolution, the recovery of dormant paleontologic life, the culturing of tissues from extinct life forms, the cloning of extinct life forms, the exploration of paleontologic host-parasite relationships, the evolution of the immune system and ancient parasites, and the role of these parasites and pathogens in the extinction of species and orders. In that newsletter Tkach outlined some of his initial ideas and announced that the First Extinct DNA conference was to be held in Bozeman, Montana, on March 9, 1983, at the Montana State University medical school.

At the last minute, the meeting site was changed and we—I and Drs. Tkach, Ron Goodwin, Roger Avery, Glen Epling, and Paul Baker—ended up at Jack Tkach's house. We taped the meeting and sent copies to Drs. Clair Folsome and Michael Rinaldi, as well as to Roberta, who had been unable to attend. If the group was small, everyone's enthusiasm certainly made up for it. The conference covered mundane but necessary topics such as terminology and moved on to the process of amberization, paleodormancy, nuclear and mitochondrial paleo-DNA survival, amber permeability, and dating.

The predominant discussion, however, centered on the problem of establishing the authenticity of the bacterial cultures isolated from different amber sources in my lab. The consensus was that all bacterial work done with amber would be subjected to criticism from two disparate camps: one side would say that bacteria are ubiquitous and the possibility of any isolated bacteria being a contaminant could never be discounted; therefore, you could never prove that what you thought was an ancient bacterium

Participants of the first Extinct DNA Conference held in Bozeman, Montana, on March 9, 1983. From left to right: Roger Avery, Paul Baker, Glen Epling, John Tkach, George Poinar, and Ron Goodwin.
(Photograph by Karen Tkach)

really was ancient. The other side would be concerned with the possible pathogenicity of any isolated bacteria and public safety. Regardless of the ultimate publishability of the results, the group members were interested enough to continue research on the amber-isolated bacteria (using the proper precautions, of course). Some details of this historic meeting were reported in the second issue of the *Extinct DNA Newsletter* in March 1983, which was circulated to some 30 to 40 people. Also included were notes on the techniques used in isolating the bacteria, the sources of the amber, the initial characterization of the colonies done by Gerard Thomas, and Tkach's idea on obtaining dinosaur DNA from mosquitoes in amber.

In the second edition of the *Extinct DNA Newsletter*, in March 1983, Jack Tkach wrote (pp. 8–9)

Somewhere there may be a mosquito that fed on a dinosaur and got preserved in amber. If one could recover a white blood cell of a dinosaur from the stomach of a mosquito, he might be able to transplant it into an enucleated egg and grow dinosaur tissue culture or ultimately a dinosaur. If the genome is partially destroyed, it might be possible to complement it by using several nuclei.

Those were exciting times. A group of people were gathered together who believed in the feasibility of amber-derived DNA research. Many others thought us fools, or worse, and some shared our views but were afraid to join us. But it was, at the very least, a beginning—and look where it has led us.

The study group, which eventually numbered between fifteen and twenty individuals, functioned as a loose coalition of scientists interested in different aspects of paleo-DNA. In April 1983 a third newsletter, including scanning and transmission electron micrographs of the bacteria isolated from Mexican amber taken by Dr. Epling and Andy Blixt, was distributed. It was the final issue.

Nearly a decade later, in April 1992, the first issue of the *Ancient DNA Newsletter* appeared, edited by Drs. Robert Wayne and Alan Cooper. Its purpose is to "communicate advances in laboratory procedures and data analysis that are relevant to extraction and characterization of DNA from ancient material." The last few pages display an ancient DNA reference list that chronicles the birth and development of this new field. Before 1984, only two papers are listed. Then, in 1984, there was a burst of activity—four papers. By 1992, almost one hundred papers had been published. Perhaps our Extinct DNA Study Group was before its time . . . but not by much.

13 Western European Collections

Amber is a household word in many parts of Europe because of its influence in the Baltic region and its extensive historical and cultural ramifications. Museums always have bits and pieces of amber tucked away, and wonderful, intricate carvings are often on display. Stores offer amber jewelry, and pipes with amber carvings and meerschaum bowls grace smoke-shop shelves. Everyone's mother or grandmother owns an amber necklace or brooch that has been in the family for years. Indeed, it would be difficult to visit Europe and not be exposed to amber somewhere. Whenever we travel to Europe, for whatever purpose, we always plan to visit some of the many amber collections there. And, of course, when I was invited to speak at the International Symposium for Biological Plant and Health Protection in Mainz, Germany, in autumn 1984, I planned visits to colleagues who were studying amber fossils.

The first amber researcher we visited worked in Basel, in northern Switzerland, where France, Germany, and Switzerland meet. This marvelous medieval city, first called Basilia around A.D. 374, is bisected by the Rhine and linked by six bridges that cross it. The old town, resplendent with antique structures, is dominated by the Munster. This cathedral of Romanesque and Gothic architecture is built of red sandstone and capped with a diamond-patterned brightly colored tile roof. Erasmus lies buried there. Close by is the Rathouse, or town hall. Constructed in the 1500s, this building sports a large decorative clock in the facade, adorned by statues and surrounded by colorful frescoes on the adjacent walls and extending into the courtyard. In the front of the Rathouse is an open-air market with flower stalls and fruit and vegetable stands. The smell of roasting chestnuts

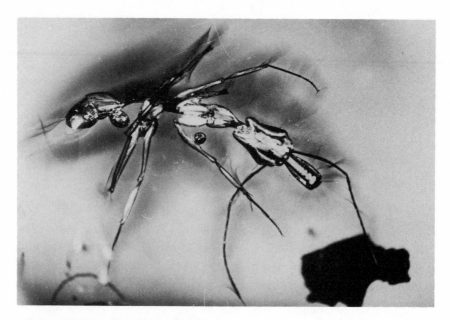

An Anochetus *ant in Dominican amber.*

permeates the square. In the midst of this venerable section of town sit the Zoological Institute, the Natural History Museum, and the Tropical Institute.

Waiting for us at the Zoological Institute was Dr. Cesare Baroni Urbani, an expert ant taxonomist. An energetic man, he was the first to describe species of ants found in the amber from the Dominican Republic. One especially impressive study by Dr. Baroni Urbani was his description of the ant *Leptomyrmex neotropicus.* Today this genus of ants is restricted to New Guinea, New Caledonia, and the east coast of Australia. He suggested that a cosmopolitan distribution of *Leptomyrmex* in the past, followed by a later contraction, would explain its present-day distribution. This theory was supported by the finding of a *Leptomyrmex*-like ant, *Leptomyrmula maravignae,* in Sicilian amber. Behaviorally, *Leptomyrmex* is an interesting group of ants that do not excavate nests but use the empty burrows of small animals. They walk in an odd position, with the abdomen pulled forward over the thorax, and the queens are wingless. Dr. Baroni Urbani identified a specialized worker subcaste of *Leptomyrmex* with a swollen abdomen in a piece of Dominican amber that I had acquired. The swollen abdomen condition is an adaptation that allows the workers to ingest and retain liquids in dry climates. This discovery provides evidence that the amber forest, 25 to 40 million years ago, probably had a dry period necessitating water transport and retention in this manner.

Dr. Baroni Urbani also described the first fossil garden ant, *Trachymyrmex primaevus,* from Dominican amber. These garden ants today cultivate fungi on pieces of leaves or other debris they carry back to the nest. The fungi serve as a food source for young larvae. Some ants he described have present-day descendants still found in the Caribbean. Today small colonies of ants that are related to the Dominican amber ant, *Anochetus corayi,* live in the soil and sometimes in trees while foraging on ground litter.

Dr. Baroni Urbani had purchased for the museum a large collection of ants from the Dominican Republic, and he and his students spent many years studying them. These fossil ants represent a window into the past ecology of the amber forest. Putting the pieces of knowledge garnered by Dr. Baroni Urbani together with those from many other Dominican amber experts throughout the world will eventually allow us to understand more about the conditions associated with that ancient forest.

We stayed in the guest house of the institute, located next to the Middle Rhinebridge, during our visit. From the window we saw huge barges plying their way along the river loaded with unknown cargoes and headed for mysterious destinations. Smaller and faster boats darted around them like bits of leaves borne on the current around a steadfast rock. The traffic along the river has continued for centuries; only the methods of propulsion have changed. Staying in that venerable guest house, I could easily imagine what it had been like to live in Basel 300 or 400 years ago and look out that same window onto the Rhine. Like amber, it was a window to the past.

Our next amber destination was Stuttgart, and we decided to drive there. The road led us through Colmar, in France, where we stayed the night. The cultural flavor of Colmar is distinctively different from Basel, yet it is only a short distance away. Alsatian Renaissance houses, with heavy timbers patterning their white plaster exteriors, rise three to five stories above narrow cobblestone roads in the old town. Window boxes filled with bright red geraniums are seen everywhere. Often the steep tiled roofs are patterned, and carved wooden doors lie in arched doorways.

Continuing onward to Stuttgart, the road passed through the Schwarzwald, or Black Forest, in southwestern Germany. This area, the source of the Danube and Neckar Rivers, is generally mountainous. The road curved through oak and beech forests, and the slopes at higher elevations were covered with fir trees. Shops along the way offered cuckoo clocks for sale, a regional specialty. Rathskellers advertised sauerkraut, bratwurst, Thuringer Kloosse (potato dumplings) for the hungry, all to be washed down with thick, dark German beer.

In Stuttgart we walked through the city center, which was almost completely destroyed in World War II and has since been rebuilt. The Rosenstein Castle is now the Natural History Museum, but we were informed that Dr. Dieter Schlee, the amber specialist we wanted to meet, was housed at the "new" building. We set off looking for this nebulous building and, on our way, passed stands of large sycamore trees planted in a central park. Their leaves were being lifted by the wind and carried across the shortly cropped grass to float on a shallow pond filled with mallards and coots.

Just as we were about to give up our search, a man—Schlee, it turned out—called to us from the window of an ultramodern flat building with bold blue trim. He met us in the vestibule and took us to the room where the amber collection was housed. The museum at that time had about 4,600 fossil amber pieces from the Dominican Republic, the most extensive collection in Europe. Some pieces were fairly large and others contained masses of insects, one with up to 2,000 ants. Also part of the collection were 2,500 pieces of Baltic amber, plus amber and copal from many other areas of the world. Dr. Schlee was meticulous in his attention to detail, and the fossils were all in individual plastic boxes with assigned acquisition numbers—very impressively organized. An amber display was in the process of being arranged for the Natural History Museum.

Although the Stuttgart collection was the last scientifically interesting amber we would see on this trip, amber jewelry was available everywhere we went in Germany. This has been true in all the European cities we have visited.

Scandinavian amber carvings from 5,000 B.C.
(National Museum, Copenhagen)

We have traveled to Scandinavia several times over the past fifteen years, usually to visit the famous amber collections in Denmark. When I first visited the Zoological Museum in Copenhagen in December 1978, Sven Larsson, who had just finished writing his book on Baltic amber, was recovering from a stroke and I spoke only briefly to him by telephone. Dr. S. L. Tuxen had taken Larsson's place as director of the famous amber collection at the Zoological Museum. We examined a few of the more interesting of the some 7,600 Baltic amber specimens. Tuxen told me that he had established the collection by purchasing amber from Danish dealers

Seventeenth-century German amber mug.
(Rosenborg Castle, Copenhagen)

Large displays of amber decorate this store wall in Copenhagen.

after the Königsberg amber collection was destroyed in the war. Ten years later, when we returned to Copenhagen, both Larsson and Tuxen were dead, but the amber collection was still there. We visited with Dr. Olie Heie, an aphid specialist who was describing a Dominican amber aphid from our collection.

The Rosenborg Castle in Copenhagen boasts a variety of wonderful historical amber carvings from the collections of Danish kings. These beautiful carvings represent the work of several artisans, one of the more famous being Lorenz Spengler, a German malacologist, who carved both amber and ivory between 1700 and 1750. His carving of a large amber chandelier is considered the most valuable amber carving in the castle. Other items made from amber include a mirror, chess set, large chest, small chandelier, and mug.

Similar carvings of amber mugs and eating-utensil handles occur in the Nordic Museum in Stockholm. Most of the mugs were made by Georg Schreiber of Königsberg around 1630. A larger collection of carved amber can be found in the Palace Museum (Kings Castle), which displays amber plates, pitchers, mugs, and bowls. For those interested in historical smoking items, the Tobacco Museum in Stockholm contains amber pipes and cigar and cigarette stems dating from 1800 to 1900.

There are many more amber collections, both scientific and historical, in Europe and Great Britain, than we could visit. We eagerly await the opportunity to see some of these others in the future.

14   Amber in the Caribbean

Some 40 to 60 million years ago, the ancient landmass that formed an archipelagic bridge between what is now North and South America began to move to the east. It was a slow process, accompanied by sporadic volcanic eruptions along the western border of this moving island as the Cocos and Caribbean continental plates, driven by powerful forces beneath the earth's crust, jostled for space.

At the beginning of this journey, rain forests containing now extinct resin-producing algarroba (*Hymenaea*) trees stretched from Mexico across the archipelagic bridge into South America and onto the then adjacent continent of Africa. No one was present to record how long it took this archipelagic bridge, now known as the Proto-Greater Antilles, to slip out of its position like a piece from a jigsaw puzzle and gradually shift over to the middle of the Caribbean Sea, nor exactly when it broke up into the land portions we now call Cuba, Hispaniola, Jamaica, and Puerto Rico.

But the events at various stages of this scenario *were* recorded—in amber. The algarroba trees on this moving island continuously produced resin that in turn perpetually trapped animal and plant life. The resin began to polymerize and harden as soon as it was exposed to the atmosphere and rapidly turned into a subfossil stage known as copal. Clumps of the material fell to the ground and were covered, first by decaying vegetation and later by soil. Water from torrential rains eroded the land and washed the fossilized resin into low-lying areas that were eventually inundated by seawater. The resin transformed into what we know today as amber, with its typical characteristics of hardness, density, and melting point. Slowly the amber was deposited in layers of sediment formed in the shallow sea bottom. These amber-containing silt beds eventually changed into limestone and sandstone strata, to be uplifted later by mountain-forming processes, which carried the enclosed amber back above sea level in a land now known as the Dominican Republic. Although the landscape was now different from that in which the resin had formed, the amber held clues to the type of life forms that were present in the original ancient forest.

Among the many life forms were stingless bees, now unknown in Hispaniola, which had swarmed over the tropical flowers and collected resin for construction of their nests. Huge termites, now extinct from this part of the world, had skeletonized the dead and dying trunks of the forest trees. Large *Leptomyrmex* ants, now restricted to the Australian region, crisscrossed the forest floor foraging for food. Lizards scampered up and down the tree branches, occasionally meeting ferocious-looking wind scorpions, and always on the lookout for predatory birds and mammals.

All of this can be construed from the amber fossils, allowing us to reconstruct that early landscape. Even complete fossils are not necessary in the case of mammals and birds; a group of hairs, or a feather alone, is sufficient. From what the amber holds, we see that this floating landmass that eventually formed the Greater Antilles was a veritable zoological and botanical garden, containing extinct life forms from a now vanished world. The resin preserved these forms so quickly, so completely, and so gently that many were caught in behavioral acts or symbiotic relationships that could never be preserved with other types of fossilization.

Is it any wonder that the island of Hispaniola has tremendous drawing powers for anyone interested in amber, particularly fossiliferous amber? Amber fossils found in the Dominican Republic are of the very best quality, and eventually anyone working with them would feel compelled to visit their source. I was particularly interested in going there to collect, for research, amber from specific mine sites. When buying amber from dealers, you are never sure from which mine the amber came. The different mines in the Dominican Republic contain amber that appears to have formed in diverse time periods. For this reason, I wanted to obtain samples of amber and its surrounding rock matrix from as many mines as possible, in order to date the amber using modern techniques. The surrounding rock matrix could be examined for foraminifera (shells of protozoa) and nannofossils (algal deposits), as was traditionally done for obtaining indirect ages of amber. The amber could be analyzed by nuclear magnetic resonance (NMR) at Northwestern University under the guidance of Dr. Joseph Lambert. The results of these procedures, when compared, would provide a better understanding of the age and botanical source of the amber from the various mines, and this could then be correlated with the types of fossils originating from each mine.

In the summer of 1986, I learned of a means to accomplish this project. The University of California Research Expeditions Programs (UREP) puts people from all walks of life together with university faculty in need of assistance for research expeditions throughout the world. The participants pay a tax-deductible contribution to help subsidize the research and become volunteer workers on the project. The projects are diverse—ranging from studies of animal behavior to marine ecology. Roberta and I decided that this would be the perfect opportunity to visit the amber mines and obtain the samples needed for this study. So, in the summer of 1987, our expedition "Forever Amber: Entomology's Petrified Past" left for the Dominican Republic.

We arrived in Santo Domingo at Las Americas airport around eight o'clock in the evening. After the cool, foggy July days of the San Francisco Bay Area, the heat and humidity that assaulted us as we stepped from the plane was oppressive. By the time we had passed through customs, beads of sweat were joining on our backs to form slow-moving rivulets that trickled downward. Luckily, our liaison in the Dominican Republic, Jake Brodzinsky, was waiting for us with his broad welcoming smile and hearty handshake. Mr. Brodzinsky, a retired American living in Santo Domingo, is an expert on Dominican amber and has collected and sold to museums some of the very best fossil insect collections in the world. He taught himself entomology in order to identify the inclusions, and he does a marvelous job spotting unique and rare specimens.

He and his wife, Marianella, generously opened their home to me, Roberta, and the other seven expedition participants (four women and three men), and it was there that our group met. Although from diverse backgrounds (including teachers, engineers, and a retired Boy Scout executive), the people who joined us were an enthusiastic and agreeable group who worked very hard to make our expedition the success that it became.

Our first day was spent at the Museo Nacional de Historia Natural arranging official sanction for our research from the director, Dr. Lambertus. A technician from the museum accompanied and assisted us throughout the trip. We were told some of the history of the Dominican Republic, including the fact that amber was used in pre-Columbian times by the Taino Indians as earplugs. The Taino, a branch of the Arawak tribes of South America, were artisans and craftsmen who wove baskets, produced pottery, and made gold masks and necklaces. When Christopher Columbus arrived in December 1492 on the island of Quisqueya, as the Indians then called the Dominican Republic, he named it Espaniola. He made contact with the unsuspecting natives and was given gifts of amber, feathered garments, and gold. The amber was merely mentioned in his journal in passing, but the promise of gold led him to establish a colony in the Dominican Republic. The Spanish colonists enslaved and mistreated the Indians. That the first colony was subsequently destroyed when Columbus sailed back to Spain is not surprising. The next year, Columbus returned and set up a second settlement at Isabela. This was later moved by his brother, Bartolomew Columbus, to the site of present-day Santo Domingo in 1496. The Europeans continued to decimate the native population and, sadly, the Taino no longer exist. Believed to number at least a million in 1492, the Taino, by 1548, were reduced to only 500, victims of slavery and introduced diseases. Africans were brought to the Dominican Republic in 1511 to replace the declining native slave pool.

On the second day of the expedition, we loaded into our van and set out to establish a base camp on the outskirts of Puerto Plata, on the north coast. We drove through the Cibao, a large agricultural valley set between two mountain ranges, the Cordillera Central, and the Cordillera Septentrional. The fields were planted with rice and coffee bushes, which were shaded from the harsh sun by introduced *Erythrina* trees. Banana plants dotted the landscape. Nestled along the road and tucked into the hills were small homes with weathered wood sides, sagging porches, and tin roofs. Laundry hung along the fences, spots of color against the dense surrounding tropical greenery. Palm trees with their crowns of fronds stood above it all like anemones in a sea of green. As we approached Puerto Plata, the sugarcane fields stretched out to the horizon, blanketing the undulating hills. We located the house where we would be staying, unloaded our luggage, and with much anticipation set out for our first amber mine—La Toca.

La Toca mine, located in the Cordillera Septentrional mountain range at an elevation of about 3,700 feet, is particularly well known for its hard, clear amber and the vertebrate fossils found in it. Amber here ranges from an estimated 30 to 40 million years old. There are actually about 200 mines at La Toca, all dug into the mountain, but only about five sites were actually being worked. The only way to reach them was to walk single file along narrow paths that led through the lush subtropical foliage and then to climb up the steep mountainside and follow along the ridges to the mine shafts. The mouths of the mines were bordered by scanty shelves around which were dumped the rocky tailings, dragged from the shaft and dumped over the precipice like the droppings a huge rodent might push from its burrow. Dark tunnels extending as far as 600 feet into the hillside, the mines are precariously supported here and there with timbers and tree trunks. The shaft is only about three to four feet in diameter, requiring you to crawl along on your hands and knees (or sometimes on your stomach) as you follow the dark gray sandstone layer of earth in which amber is found. The shaft slopes up and down, and the low areas often are full of standing water, which miners must crawl through to reach the area being worked. Miners carry with them only a candle for illumination, a hammer and chisel to extract the amber, and a sack for amber chunks. The work is strenuous and can be dangerous, especially during the rainy season.

The miners were very friendly and generous. They took those of us who wanted to go into the shaft to view the amber in situ. They allowed us to collect amber samples and the surrounding sandstone, to measure and to photograph. All the miners at every site we subsequently visited had the same friendly, open nature, for which we were most grateful. Without their cooperation, our survey would have been difficult to complete.

Most of the day was spent slipping and sliding along the cliff face or crawling into the dark, damp tunnels to collect samples from several sites in the area. In the late afternoon, we climbed down to the cluster of gray wood houses and thatch-covered huts where a fossil collector lived. He brought out onto the porch rough amber soaking in a container of cooking oil (to make the amber more transparent), and everyone gathered around

Homes of amber miners in the Dominican Republic.

A Dominican amber miner following the amber vein.

as I sorted through the pieces and began to look at the fossils. Our group was the center of attention, and shy children stood in the doorways of the surrounding houses, watching us with curious brown eyes. Chickens scratched in the dirt, and flies buzzed around us looking for a likely place to land. After a while the children came out of the houses, and some resumed their play while others edged closer, interested in the proceedings on the porch. We purchased some thirty pieces, and left tired, hot, and very dirty—but satisfied. The beach at Puerto Plata curls along the Atlantic and is one of the most attractive in the Dominican Republic. The sands are white and dotted with shells, and palms fringe the beach edge. During the day it is a busy tourist spot, but at sunset it is a pleasant and quiet place to relax, with the waves lapping the sand and the ocean breeze rustling in the trees. There we planned the following day's trip.

The next morning we left for Santiago to visit an amber dealer, Ramon Martinez, known as Rubio by almost everyone. Santiago de los Treinta Caballeros is a large city of about 500,000 people located in the Cibao valley. It is a mixture of the old and new—mules occupy the streets with cars, and disco music competes with *merengue.* Rubio took us to La Buscara mine, which, like the majority of amber mines, is located in the El Mamey formation of the Cordillera Septentrional. The amber found in the rock is hard, clear, and yellow, and that found in the soil is red. The amber-containing veins follow along a streambed, and many of the deep open pits from which the amber comes eventually collapse from the saturated soil. Here the miner's implements are the shovel, the pickax, and the ubiquitous machete. The fossils we collected and purchased there came from along the streambed. Near La Buscara are the Palo Alto, Los Aguitos, and Los Higos mines, which we also visited. The Palo Alto mines offer several veins of amber on two to three different levels. We stopped at the lowest and spread out like foraging ants, crawling up the slag heap of an abandoned mine, shifting through the debris, and picking up pieces of amber that had been discarded. With eight to ten pairs of hands, we collected a fair number of small chips and pieces. Farther up the mountain face, another mine site was being worked—a deep pit dug into the steep hillside. We purchased and obtained samples from the miners there.

Up the road at Los Aguitos, an eerie fog shrouded the mountainside. In the insular quiet, the dampness settled on our sweat-drenched clothing—

the dramatic change in temperature, from the hot slag heap to the fog, thoroughly chilling us. Of the eight or ten mines at this site, we were able to sample only one from a short tunnel. After purchasing amber from the miners at these and other sites, we returned to Puerto Plata feeling rather stiff and sore. Two days of crawling, climbing, and bending were beginning to take their toll.

Early the next morning we again met Rubio and went with him to the Palo Quemado sites. We trooped along a stream and then up the mountain. The mine shaft went into the mountain about thirty or fifty feet. I crawled in, following the thick blue vein, until I encountered some men working. They lay propped up on their sides or crouched in the flickering light of two candles, striking with their hammers and chisels the hard carbonized vein of rock, chipping out the amber. The amber in this mine is about 20 to 23 million years old and yellow to red in color. This, we were told, was the first Dominican amber mine. It opened about fifty years ago, with its peak production about eighteen years ago. On our return, Rubio took us to a planted algarroba tree just outside Santiago. I had explained to the group that tests had shown that Dominican amber had been produced by ancient algarroba trees belonging to the genus *Hymenaea* of the legume family. The *Hymenaea* trees that produced the amber were now extinct, but we could find their leaves and flower parts in the amber. We were standing in front of a modern-day *Hymenaea,* a descendant of the amber tree. There were definite similarities between the two trees. In the modern species, petals and stamens became detached soon after the flowers opened. We can imagine that the same occurred with the ancient tree, since many detached petals and stamens occur in Dominican amber. Differences in the shape of the petals clearly separate the two plants, however. The petal shape of the extinct amber tree most closely resembles the petals of a primitive *Hymenaea* tree now found in East Africa; this shows an ancient connection between certain floral groups in the West Indies and Africa.

On Friday, we left the amber mines and visited amber processing factories, museums, and stores in Puerto Plata. Puerto Plata, the silver harbor, was discovered by Christopher Columbus. The port, where cruise ships now dock, was established there in the 1500s. Charming Victorian gingerbread and bright colors decorate many of the older houses in the city. Many shops in this tourist area sell amber to the cities' visitors. In the

The Costas next to their amber museum in the Dominican Republic.

morning we visited the shop and work rooms of Ramon Ortiz. He gave us a tour of his amber-working facilities, where many young people ground, shaped, and polished amber for jewelry and fossils. He owned a very nice collection of fossils, including a lizard and a pair of mating fulgorid flies.

Also in Puerto Plata was the Fundacion Dominicana de Desarrolor, run by Heinz Meder. This workshop had been built with funds supplied by the World Bank. Here students were taught how to make handicrafts and various artworks out of leather, wood, amber, and larimar. The students working with amber used a wet grinding process to protect their lungs from the insidious amber dust, something not practiced in other amber factories.

The Amber Museum in Puerto Plata is located near the plaza in a beautiful wood structure. Owned and operated by Didi and Aldo Costa, it has one of the largest displays of fossiliferous amber in the Dominican Republic. The Costas opened the museum in 1982 and since then have added more exhibits. Our group explored the gift shop downstairs while I talked with Aldo and Didi about their collection. Just as we were ready to look at the museum, the electricity went out. This was common in the Dominican Republic at the time of our visit. Electrical power was tenuous at best, and wondering whether it would stay on, along with worrying about the water, were two nuisances we constantly had to deal with. Aldo Costa was up to the challenge, though, and he led us to the museum, equipped with flashlights. The lights soon went on again and we were able to enjoy the very artistic and creative displays.

That afternoon we traveled along a road that passed through extensive sugarcane fields, and by small, colorful roadside stands stocked with tourist goods, to the city of Luperón. We were told that white amber we were seeking there had been dumped by fishing boats from the Bahamas. By searching around, we were able to locate a man who took us to a site where he collected it.

I had been hunting for Perubia, a powder made of resin and commonly used for a variety of ailments, including headaches and muscle pains. This is one of many medicines produced throughout the world, based on the attributed curative powers of resins. Whenever I come across them I purchase a bottle for future reference. We eventually found the Perubia in a crowded drugstore in Luperón.

Since a good part of the afternoon remained, we decided to proceed to El Higo, where we had heard there were two places to find amber. The dirt road leading to El Higo was rough, and we had to ford two rivers and ease our van over bumps and around potholes in the road. From our parking site, we walked about forty-five minutes to the El Higo River and along a large sliced area in the bank. There, at last, appeared a vein of soft blue earth, where we found red amber, oxidized from lying in the oxygen-rich soil. The trip back to the car led us through lush fields of tapioca and taro, and finally through a seemingly endless cornfield. The cornstalks towered above us as we trudged along the rows to the sounds of insects humming and the squish of our boots sucking in the wet, steamy fields.

Would you believe that here, in the middle of nowhere, we were sold some plastic as amber? This was the one spot, so isolated from tourism, that we never would have expected to find fakes, but there they were. You can never be too careful.

On Saturday we left for the southern part of the island, heading out toward Santiago and then to La Vega, the gateway to the Cordillera Central

Region. This area is almost a wilderness, with the nation's largest mountain range, cooler weather, and numerous waterfalls. As we climbed into the mountains, the views were spectacular but the roads were very bad. The palm trees and lush tropical greenery gave way to a five-needled pine, and after emerging from an extensive previously burned area, we continued into a fairly dense forest of pine mixed with broad-leafed Cecropia trees dotted with beds of yellow *Allamanda* flowers and even areas of *Eucalyptus.* We passed through Jarabocoa, from where people can embark to climb Pico Duarte, at 10,417 feet the highest mountain on the island.

After several days of sightseeing, we drove on to Boca Chica. Unfortunately, a general strike was scheduled for the next day and our contacts warned us that the situation could become ugly, especially for North Americans. The strike, organized by a committee comprising the seven main labor unions, was called to demand a 62 percent hike in the minimum wage for public employees, which ranged at that time from $70 to $114 per month. The workers also complained about the drop in sugar prices (which account for 34 percent of export earnings) and the rising national debt. We were urged to stay in the house, with the doors locked and the curtains drawn—and we really were afraid.

That night the women in our group slept in the downstairs living area; the men took one upstairs bedroom, and Roberta and I occupied the other. We all awoke in the middle of the night to the bloodcurdling screams of one of the women, followed by the panicked screeches of others. We jumped out of bed and flicked the light switch, but the electricity was off. Roberta found a flashlight and rushed down the stairs as the shrieks diminished. The men searched for weapons. Were we being attacked? Was an intruder stalking about? Adrenaline pumping, we stumbled down the dark steps, shouting and groping for our way. We arrived to find everyone huddled on their makeshift beds in various stages of hysteria. But there were no attackers—it was a false alarm. Apparently, one woman had been snoring and another had gotten up to shake her awake. The snorer awoke, saw a dim figure looming over her, and panicked. Her screams had set off a chain reaction. The experience shook us so much that many of us couldn't sleep for the rest of the night.

The next day a subdued group clustered around the radio listening to the strike news. Nine homemade bombs went off in various cities, includ-

Amber Odyssey *

A plant-feeding lace bug (Tingidae: Hemiptera), with beautifully intricate wings.

* *Unless otherwise noted, all color images are of Dominican amber, ranging from 25–40 million years old. All photographs by George Poinar.*

A scorpion (Buthidae: Scorpiones) is poised in a defensive position.

A planthopper (Fulgoroidea: Homoptera) with a unique head ornamentation.

A female mosquito (Culicidae: Diptera). What was its last victim?

A biting midge (Ceratopogonidae: Diptera) in Canadian Cretaceous amber, 70–80 million years old. A parasitic mite is still attached to the back of the midge.

A female fly (Diptera) that deposited an egg immediately after becoming entrapped in the sticky resin.

A plant-feeding weevil (Curculionoidea: Coleoptera) that was probably searching for food.

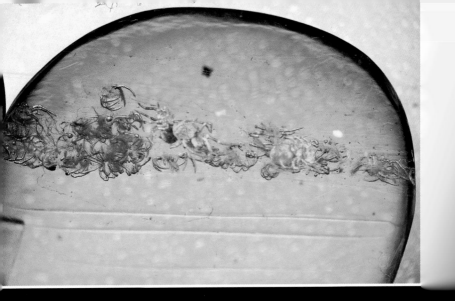

Remains of a group of spiders on strands of silk.

An ant (Formicidae: Hymenoptera) struggling to free itself from a spider web.

A large robust spider (Araneae: Arachnida,

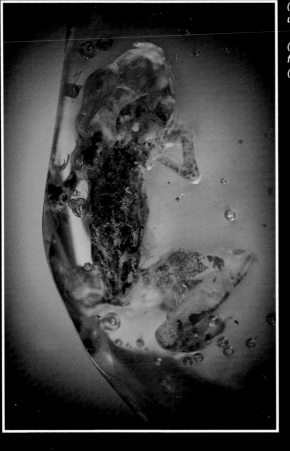

(Left) A young frog made an unfortunate leap.

(Below) The head and front foot of a female gecko (Gekkonidae: Reptilia).

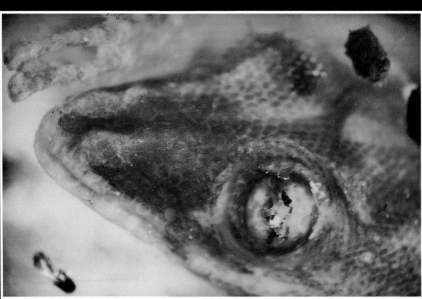

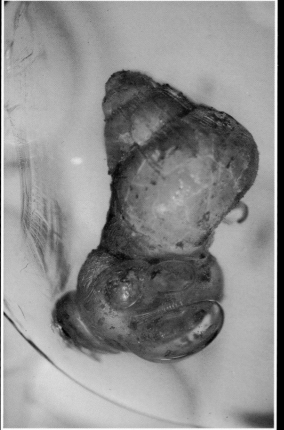

(Top) A moth (Lepidoptera) with its proboscis extended.

(Left) A land snail (Prosobranchia: Gastropoda) with its soft tissues still remaining.

right) A predatory ant bug
Reduviidae: Hemiptera)
covered with stiff protective
hairs.

(Below) A male twisted wing
insect (Strepsiptera) searching
for its wingless mate.

A stamen of the amber producing tree (Hymenaea protera). Pollen grains have fallen out of the mature anther.

A mature flower of the amber producing plant (Hymenaea protera) with its developing seed pod.

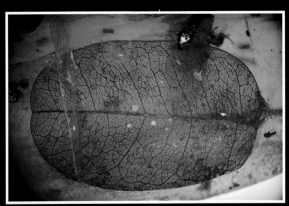

A partially cleared leaf from a legume tree.

The delicate cap of a mushroom (Coprinites dominicana). A small mite is adjacent to the rim of the cap. It didn't jump off quickly enough!

ing one at the Parque Independencia in Santo Domingo. We heard that someone had been killed by a policeman in Puerto Plata and that truckloads of policemen were shooting into crowds in the capital.

By late afternoon we were encouraged enough by the local peacefulness to venture out onto the nearly deserted beach. I swam out to the reef and observed the brightly colored fish and strange swaying plants. Seaweeds looked like mushrooms, and zebra fish and eel-like bottom fish darted among coral stands. Sand dabs and small crabs dotted the rocky bottom. The surprisingly cool waters of the Caribbean eased the tension of the previous twenty-four hours, and the group felt revived.

I awoke early the next morning to a light show of sunbeams, playing across the bedroom walls, marred by a cacophony of asses braying, roosters crowing, dogs barking, and cats caterwauling. This was accompanied by the chirping of birds and crickets and the croaking of frogs. Along the road outside, motorbikes zoomed by and cars honked—the strike was over.

Our first stop that day was at the workshop of Cesare, an amber carver in Santo Domingo producing some fine animal carvings. He told us that he preferred to work with the amber from the El Valle mine because it wasn't as prone to fractures as other ambers. We set off for that mine, driving through San Pedro de Marco, a famous baseball town. Of the many Dominicans who play baseball in the United States, about half come from this seaport town located at the mouth of the Río Higuamo. The village is surrounded by sugarcane fields, the major source of income for its citizens. From there we drove northward through the town of Hato Mayor and reached a unique settlement about seven kilometers before El Valle in the Cordillera Oriental. Here we met a singular black Dominican, Mary Johnson. Ninety years old, she was descended from ancestors with names such as Green, Kelly, Jones, Anderson, Shepard, and Johnson, who had immigrated to the Dominican Republic from Philadelphia in the early 1800s. She explained to us that her people had kept to themselves and still continued to speak very "North American" English. It was on her land that one of the El Valle mines was located. The miners in that region look for a white ash in the soil, which indicates that amber is not far away. We found the mine, but the soil had caved in around it. By searching around the rocks on the slag pile, we were able to find some small pieces of amber. The second El Valle mine was known as Ya Nigua, after the river that flows near it.

We walked about two kilometers to the river and discovered an assortment of fossilized shells along the banks, the likes of which we had not encountered in other parts of the Dominican Republic. There were long spiral turret shells, stocky dogwinkles, crown conchs and slender volutes. One group member found a large, perfect conch shell weighing about twenty pounds. We later learned that these shells came from Eocene deposits some 40 million years old. The trail continued through the dense tropical undergrowth along the streambed and on the edge of the surrounding fields of coffee and corn.

As we trudged through the undergrowth, I noticed large, unsightly brown deposits on the forks of small trees. My curiosity eventually forced me to stop and start chipping away part of one with my geologist's hammer. The surface was quite hard and required a few good blows for me to dislodge a portion. Just as I saw that the inside was filled with hollow tunnels, a number of small insects suddenly appeared along the broken portion. These strange insects had elongated heads that formed a snout or tube at the end. I recognized them at once from specimens encountered in Dominican amber: I had broken into an arboral termite nest and was being challenged by the nasute, or snout-containing soldiers. Instead of defending their nest by biting, these nasutes squirt a strong-smelling, sticky secretion out through their snout. The substance repels other small insects and is distasteful to anteaters. The hard wall of the nest is made from wood that is chewed by the workers and cemented together with fecal glue. Both their nest construction and their means of defense clearly have been essential to their survival for some 25 to 40 million years.

A much larger termite known as *Mastotermes* turns up frequently in Dominican amber, yet is absent not only from the Dominican Republic but from all of the New World. Only a single species lives today, and its range is restricted to northern Australia and some surrounding territories. Why its range shrank so during the past 25 to 40 million years is a mystery. It does construct strong nests that often protrude from the soil surface, and the soldiers possess slashing mandibles that tear like a pair of scissors into any attacker. *Mastotermes* soldiers augment this first line of defense with the release of gluelike head secretions that entangle an attacker. These abilities sound as if they would be key to survival, but obviously they weren't good enough for the species that used to thrive in the neotropics.

We followed the river to three more mines, one full of water. Miners in the other two were down two to three meters into the typical blue layer of earth in which amber is found. We purchased amber from each site and collected oxidized red amber from the overlying soil. The amber from El Valle occurs in all colors, even pink and blue. It varies from hard to soft and has yielded many pseudoscorpions and some vertebrates, as well as many insect fossils. The El Valle amber was estimated to be 20 to 30 million years old.

We visited our final mine near the town of Bayaguana, located in the foothills in the western part of the country. We heard that amber could be found at two sites—Comatillo and Sierra de Agua, and in Bayaguana we found a guide who volunteered to take us to both. At Comatillo, amber was being sold in a store for 260 pesos per pound; miners who sold it to the proprietor received only about 60 pesos per pound. We bought some pieces, and the shopkeeper gladly talked to us about the mines. He indicated that amber could be found in four layers of soil of varying thickness: a top white layer, followed by a yellow layer, a lower black layer, and finally the lowest layer, with a slight bluish tinge.

On leaving the store we set out in search of the mine and, on our way, encountered some Haitians also searching for amber. Haitians, who form the largest foreign minority group in the Dominican Republic, are hired to harvest the sugarcane and frequently work long hours below minimum wage. They live in barracks, sometimes without light, running water, or toilets. Agents, who are paid $7 to $20 per head, actively recruit them from Haiti. The Haitians we met led us to a deep mine dug into the ground where they were removing oxidized amber from the ashen white layer. They gave us samples in the rock for later analysis.

We continued hiking along the river to the next site, which consisted of a large cave with a hole in the bottom filled with water. The layers of amber-containing soil were clearly visible around the face of the cave, but one layer lay at the bottom of the water-filled pit. Our guide explained to the miners our purposes for collecting rocks from the various layers of soil and conveyed our disappointment that the final layer we were interested in sampling was submerged. To our surprise, a young boy dove into the hole and emerged after a short time triumphantly holding the rock sample we needed.

A middleman purchasing amber from a miner in the Dominican Republic.

This spontaneous behavior was characteristic of many miners we encountered; they were always willing to help. These people who bring forth from the earth the fossils we treasure lead extraordinarily hard lives by any standards. The labor is backbreaking in the hot, humid environment. They earn very little for the amber, and many middlemen take their cut before it finally reaches the collector. Just how perilous an amber miner's life can be was made clear in 1993, when three miners were killed at La Toca mine. The rainy season had arrived, and the men were pumping water out of the mine. Suddenly there was a landslide, and all three—Andres Polanco, Ramon Jaquez, and Victor Jaquez—were asphyxiated and crushed to death.

We spent the following day savoring the atmosphere of the old town in Santo Domingo and visiting people dealing in amber. Carlton Rood owned a shop on the Calle Las Damas filled with amber, some pieces containing rare and valuable fossils. Rood, an American, had retired to the Dominican Republic and, like others, had started an amber collection. He also had written the book *A Dominican Chronicle,* one of the few histories of the Dominican Republic written in English. We enjoyed the opportunity to talk with him about Columbus, Sir Francis Drake, colonization, pirates, and many other colorful aspects of Hispaniola's past.

Our final stop in the afternoon was a visit with Jim and Gloria Work. Jim, an American amber dealer who once worked in the Dominican Republic for the Peace Corps, had gathered together an extensive collection of fossils and jewelry for us to examine. He represented the final link in the chain of people who bring fossils to buyers and collectors. We had now seen all of the hands that amber fossils pass through before reaching the scientist: the miners who dig it, the processors who shape and polish it, the suppliers who may sell it in their stores in the Dominican Republic or to the foreign dealers, and the dealers who take it out of the country and sell it to stores and collectors.

On August 2 we left the Dominican Republic and considered our trip a success. No one had become seriously ill, been bitten by scorpions, or fallen down the mountainside. The strike, although threatening, had not harmed us and was offset by generous and helpful people we met along the way. Already the problems—rainstorms, rough roads, biting insects, and erratic electricity—were fading and being replaced by lasting memories of palm-lined, white sand beaches lapped by gentle waves, by lush tropical vegetation festooned with flowers and ferns, and by the songs of birds and insects. But some sad things we will never forget, such as the mercuric stream contaminated with chemicals from an upstream gold mine, a stream from which people drank and in which children played; after almost 500 years, the gold that had drawn Columbus to Hispaniola was still killing its people. Nor can we forget the children running after our van with their hands outstretched, begging for money.

Our many amber and rock specimens from the Dominican Republic were sorted through and sent out for analysis. The rough amber was polished and examined for insect inclusions. We now had amber that we knew with certainty came from particular mines because we had collected it ourselves. We had samples of the surrounding bedrock.

We combined the floral and vegetative plant fossils in amber collected on this trip with other samples and examined them over a seven-year

Amber dealer Dan McAuley holds one of the largest pieces of amber (seventeen-and-a-half pounds) from the Dominican Republic. This prize is displayed in the Ancient Stone Works shop of Mary and Jack Gentry in Santa Fe, New Mexico.
(Photograph by Pat Craig)

period for my study that described *Hymenaea protera,* the fossil algarroba tree that probably was the source of much of the amber found in the Dominican Republic. This fossil tree most closely resembles the present-day *Hymenaea verrucosa* now found in East Africa. The evidence suggests to me that the precursor of *Hymenaea protera* covered a large area that encompassed the Proto-Greater-Antilles, as well as over the then adjacent areas of East Africa and South America. As the continents began to drift apart in the late Cretaceous, these trees underwent speciation. Eventually, climatic and biological factors led to the extinction of *Hymenaea protera,* although some thirteen related species still exist. The amber tree's closest relative survived in ecological niches along the east coast of Africa, and in parts of Malagasy and the islands of Seychelles and Mauritius. Later DNA studies confirmed that the Dominican amber tree was most closely related to the only existing species in Africa.

15 Amber Down Under

The two islands forming the nation of New Zealand, together in size only about two-thirds of the area of California, provide a rare glimpse of "living fossils in the flesh." New Zealand's separation from the rest of the world occurred early, and the animal and plant life on it survived in splendid isolation when similar forms on other landmasses became extinct. Species developed slowly in this island paradise, because they could depend on a continuously stable habitat; and although many of the endemic forms did take a long time to mature, they had amazingly long life spans. Some of these relics, or living fossils, still exist today for scientists to study, although habitat destruction and the introduction of predators (rats, cats, dogs, opossums, weasels) by humans are rapidly changing the scene.

Animal and plant diversity were not great in New Zealand; imagine a land with no terrestrial mammals except for bats. But these bats, known as the New Zealand short-tailed bats, belong to a family (Mystacinidae) found nowhere else and have adapted to survival in thick forests. Although capable of flight, they have strong hind legs and feet with talons on their claws that enable them to run like squirrels up tree trunks and along the branches. Their wings fold up into skin pockets, and they have an extendable papillose or papillae-covered tongue.

Then there is the tuatara, which looks like a lizard but is the only living representative of a completely separate group, the Rhynchocephalia. All others from this group have been extinct for at least 60 million years. In fact, the tuatara is almost identical to the fossil *Homoeosaurus*, which existed 140 million years ago (Upper Jurassic) in Europe. Tuataras can live for up to one hundred years and are now restricted to offshore islands free of introduced rats, which destroy these unique reptiles.

Only three species of endemic amphibians exist in New Zealand—all frogs belonging to the archaic genus *Leiopelma*. These most primitive of all frogs have unique skeletal and muscular features. You can't hear them, because they lack both vocal sacs and eardrums and at most can utter a faint squeak. They deposit their eggs on land, and the tadpole develops in a gelatinous capsule surrounding the egg. In two of the three species, the male guards the developing eggs and the newly hatched froglets remain on the male's back for a month before striking out on their own.

New Zealand birds are no less bizarre. The kiwi, a flightless bird adapted to dense forests, comes out only at night. It sees poorly and finds food in the ground using its long sensitive bill and bristles around its mouth. The males incubate the unusually large eggs. The kiwi and the moa coexisted in New Zealand until the arrival of the Polynesians, some 1,000 years ago. Human activities eliminated the larger moas, now considered extinct. Some naturalists still hope that somewhere in the vast stretches of southern beech forests in South Island some moas may still survive. Their hopes were rekindled when the takahe, a colorful, flightless gallinule (wading bird) thought to be extinct since 1898, was rediscovered in 1948 in the beech forests.

Some New Zealand insects are also living relics. Consider the wetas, a group of large flightless insects, related to our grasshoppers, that evolved into niches that normally would have been occupied by mice and other ground rodents. These plant feeders emerge at night to browse on leaves. They have diversified into three groups—ground wetas, tree wetas, and giant wetas—each exhibiting different types of behavior, food preference, and habitat selection. These ancient insects, curiously, are themselves parasitized by another relictual group, the hairworms—a relationship that probably extends back several hundred million years. Hairworms, abundant in some parts of New Zealand, depend on wetas since they can complete their development only inside the body cavity of a living weta.

▶

The New Zealand kauri pine (Agathis australis).

Although hairworms kill their insect hosts when they emerge, a balance has been established between these two animals that does not jeopardize the existence of either one.

Archaic life forms in New Zealand are not limited to animals. A number of plant genera are unique in their own way, and one of the most interesting to us was the kauri pine, *Agathis australis.*

When beginning our amber investigations, we thought it would be interesting to find an amber-bearing region where the original tree that produced the amber was still living and producing resin. Then we would have a story spanning millions of years, laid out at our feet. Not until later did we realize that such places are extremely rare—but one of them is in New Zealand. Our appetites were first whetted when our colleagues Eder and Jim Hansen brought us some fossilized kauri pine resin from New Zealand in 1985. After examining these pieces, we decided that we must go there ourselves to collect a variety of samples.

The New Zealand kauri pine is one of the relatively few trees producing resin that can persist and change into amber. These trees belong to the family *Araucariaceae,* which today comprises only two living genera, *Agathis* and *Araucaria.* While representatives of the little-known kauri pines (genus *Agathis*) sometimes appear in botanical gardens in warmer areas of the United States, members of the genus *Araucaria* are often planted as ornamental trees. The most well-known species are the Norfolk Island pine and the monkey puzzle tree.

Both genera occur naturally today only in the Southern Hemisphere, and a greater resin production occurs in *Agathis* than in *Araucaria.*

Geological evidence indicates that during past eons, kauri pines grew over most of the land surface of New Zealand. Today, however, precious few of these magnificent trees remain, most having been cut for lumber. These trees attain a great size and age (up to seven meters in diameter, fifty meters in height, and more than 1,000 years old) and are equivalent in many respects to the redwoods of California. Thanks to the perseverance of a few environmentalists, a small area of mature kauri pines was saved as a reserve. Aside from this Waipoua Forest Sanctuary, a few smaller reserves and stands of several trees survive, all found only in the northernmost portion of the North Island (north of 38° latitude).

Remains of an ancient kauri pine forest in New Zealand.

A petrified tree from the Jurassic period, now bathed by waves where the Tasman Sea and Pacific Ocean meet, could be one of the oldest known kauri pines.

The kauri pine bears no resemblance to the pines we are used to seeing in North American forests. Kauri pines don't have needles, but thick broad leaves shaped similar to those of acacias. However, the pinelike cones show that they are true conifers. The bark of the New Zealand kauri pine is unusual in that the surface is composed of large flakes or scales that periodically fall off and are replaced by others. This is said to be how the trees rid themselves of moss, lichens, and other life forms on their trunks.

Entering the Waipoua Forest is like moving through a time warp into the past. The giant trees covered with epiphytes create the atmosphere of a prehistoric forest. The ground cover is composed in part of silver-leaf ferns, giant Dawsonia moss, kauri grass (actually a lily), huge clumps of the sharp-leafed Ghania sedge, the scented leaves of the shrubby myrtle, and yellow flowers of the groundsel. Overhead are the purple berries of the tahaire tree, the broad foliage of the dracaena-like dragon leaf, the five-fingered leaves of the false panax tree, and the crimson, fragrant flowers of the large-leaved New Zealand honeysuckle.

A number of invertebrates are also associated with the kauri pine forests, the most notable being the endangered kauri snail, a large carnivorous gastropod whose habits unfortunately offer no protection against the introduced rats. Insects whose existence depends on kauri pines include a kauri moth, several types of kauri weevils, a kauri mealybug, and a kauri thrips. The moth is so unique and primitive that it has been placed in a separate suborder of the Lepidoptera. The young larvae of this moth feed within the kauri seeds.

The atmosphere under the canopy of these ancient trees is quiet and peaceful, and as one stares at the resin running down the trunk and accumulating at the base, the question presents itself: What is the purpose of this resin? Is it just a defense against herbivorous insects and fungal pathogens, as some have suggested? More plausible is the idea that if used as chemical defense, it had to be the result of selection pressure while the tree was evolving into its present form. What was around in the Mesozoic when these trees were developing? The answer is, of course, plant-eating dinosaurs. Whether dinosaurs ate kauri leaves, or avoided them because of their resin content, may never be known, but resin production in Mesozoic conifers may well have evolved as a means of protection against herbivorous dinosaurs.

In the North Island of New Zealand, kauri pines have been growing for at least the past 40 million years. When the first humans arrived some 1,000 years ago, these Polynesians noticed that semifossilized resin, now known as copal, dotted the landscape and could be found in the soil, down to depths of five meters. The Maori used fresh resin, or *kaapia,* as a chewing gum, because it contained sugars and also cleaned their teeth. Even the harder semifossilized resin served this purpose, after the Maori softened it by mixing it with the sap of the sow thistle. The Maori also discovered that the copal would burn and used it for starting fires and providing light. They noticed that the acrid smoke killed the caterpillars of the sphinx and hawk moths that attacked their sweet potatoes, thus providing an early record of insect fumigation. Burnt resin and copal, sometimes mixed with other compounds, were used to produce the pigment for tattooing. This burnt resin was actually placed on a bone chisel, which was then forced through the surface of the skin, imparting a blue color to the bearer. Still more intriguing are the medicinal uses of copal by the Maori. Powdered copal was mixed with oil and applied to burns and wounds as an antiseptic.

After the European settlers arrived, in 1825, the Maori collected and sold copal (called "kauri gum" by the newly arrived immigrants) to the newcomers as a source of fuel. By 1835 some settlers had discovered that kauri copal could be dissolved in oil and mixed with paint, or made directly into a varnish. In fact, kauri varnish ran easier than other varnishes and produced a coating of superior hardness and brilliance. Thus copal became a marketable product, and that meant profit and employment. The Maori were the first "gum diggers," as the gatherers of kauri copal were called. They sold it to buyers for roughly five pounds per ton, and the first commercial shipment abroad left Hokianga harbor in 1836. By 1843, large quantities of copal had already been exported to England and America, where it was used for the manufacture of varnish. The surface deposit of copal was soon exhausted, as foreign markets for the resin appeared, and by 1855, workers had started digging in earnest for the eagerly sought-after material.

In 1867 some 73,000 immigrants in the North Island were looking for employment, and gum digging required little experience and, with luck, could be profitable.

A gum digger with the tools of his trade.
(Historical Northwood Bros. photograph)

Storekeepers decided to become involved and often provided "grub stakes," which consisted of a spade, a spear, billies, cooking utensils, a tent, and some food; the digger receiving these items was obliged to turn in his first month's supply to the store owner. The procedure for locating gum was rather routine: After setting up his tent, the digger would take his one-meter-length spear (a pointed metal rod with a handle at one end) and prod the ground until he struck something. With experience, he could tell

◄

A Maori head carved out of kauri copal.
(Property of Lionel Ewidnton, Christchurch, New Zealand)

A Maori scraping soil from kauri copal in New Zealand. (Historical Northwood Bros. photograph)

whether it was a rock, a piece of wood, or a lump of copal. If the latter, it was dug up with the spade. Then, to bring the highest price, the piece had to be cleaned of dirt and the crumbly oxidized surface scraped away. From 1870 to 1886, between 3,000 and 6,000 tons of copal were exported annually, representing the labor of 1,000 to 2,000 full-time diggers.

So many gum diggers were working by 1898 that regulation became necessary. Thus, in that year, the Kauri Gum Commission passed the Kauri Gum Industry Act, and some 290,000 acres of gum-digging land was set aside as Kauri Gum Reserves for the Maori. Only naturalized subjects could dig, and licenses were required.

A group of Yugoslavian diggers holding their prize of kauri copal.
(Historical Northwood Bros. photograph)

With so many looking for gum, groups and partnerships often formed to streamline the work and make conditions more sociable. James Bell, who traveled through Northland in 1909 and 1910, reported that Maori families worked together:

Here and there we came upon groups of Maoris, who greeted us cheerfully and lent a human interest to the landscape. The largest number of these natives were busily engaged in collecting kauri gum from a swampy flat which, like many others in these parts, filled a shallow depression among the hills. Unfortunately for our appetites, we came upon these Maoris in a time of excessive leanness, and all alike—men, women, children, young and old, grizzly warriors, tattooed wahines (women) and black-eyed babies—seemed busily engaged in gum-hunting. Some, bare to their waists, stood submerged to their middles in the wetter holes in the swamp, feeling for lumps of gum, which they brought up from time to time with their feet, or dived for with their hands, if neither of their limbs were successful in bringing the treasure to the surface. Others walked over the more solid parts of the swampy ground, prodding with a long iron spear. When a piece of gum was located, it was brought to the top by a long iron hook, or, if the pieces were big, it was sometimes necessary—owing to intervening roots—to dig for it.

During this period, the ranks of the diggers swelled to some 14,500, who, aside from the Maori, came from many different countries. The largest known piece of kauri gum was discovered by a digger at Mangawhore in 1898, weighed 185 pounds, and is now on exhibit in the Northern Wairoa Museum in Dargaville.

Perhaps best known—not because they were the most numerous, but because they lived and worked together and were very thorough diggers— were immigrants from Yugoslavia, also known as Dalmatians or Dallies (Dalmatia was then a province of Yugoslavia) and Austrians (at that time both Dalmatia and Croatia were part of the Austro-Hungarian Empire). Most had been farmers in Yugoslavia living under sparse conditions. A hard life, bleak prospects for the future, and a yearning to be free from Austro-Hungarian domination, especially during times of forced conscription into the empire's military forces, inspired them to leave. They began arriving in New Zealand in the 1890s, and by 1898 they numbered about 1,600. Their custom, common to many south and east European immigrants, of living frugally, working hard and long, and sticking together did not endear them to the resident New Zealand Pakehas (non-natives). But it did result in their relative success in the gumfields. With no capital, and few opportunities to work at other jobs, the Yugoslavs had little else to do but dig kauri gum.

The Yugoslavs who worked in teams methodically and systematically turned over every square foot of soil in an area, often down to six feet in depth. Their methods were so precise that Roy Wagener, in *The Desert Shore,* states that "most local gum diggers would return again and again to work over places in the hope that they may find a patch they or others had overlooked but this was a useless gesture in any land where the Dallies had worked."

The working conditions were hard, and James Bell recounts meeting a Croatian on the gumfields in 1909 who "told us how grievously scarce the kauri-gum had become, and to what extent his revenue had dwindled accordingly, though now, longing to return to his native Croatia, he worked longer hours and harder than formerly." In those days, most of the country inns were filled with their quota of Maori and Croatian gum diggers, where, at least, they could get a meal and dry their wet clothes near the kitchen stoves.

Toward the end of the peak exportation period, the scarcity of kauri gum drove people to use destructive tactics to obtain it. One such practice was to bleed kauri trees. Climbers would chop wedge-shaped incisions into the tree bark at regular intervals. Within a few months, the resin had extruded through the wounds and hardened enough to be chopped off. Then the wound would be reopened to initiate the next harvest. In 1905 the Crown Lands Board ruled the practice illegal. Violators could be fined five pounds and receive three months' imprisonment. Later, however, on the recommendation of an outside consultant who reported that if properly performed, gum bleeding would not injure the tree, the practice was reinstated. In fact, the Kauri Gum Industry Report for 1916 suggested that the forests be maintained for an annual revenue that could be obtained by gum bleeding. Eventually, tree bleeding was outlawed, but not until many trees had died from overzealous use of this practice.

After 1965 the annual exportation of copal never exceeded fifty tons. The reasons for the decline were clearly listed by J. R. Hosking as the following: (1) the high cost of kauri resin in comparison with other natural resins of the same class; (2) the difficulty of obtaining a standard product due to the presence of so many grades; (3) the appearance of synthetic resins on the market (with standard, reproducible properties); (4) an insecure future supply, which forced manufacturers to turn to synthetics; and (5) the demand for quick-drying finishes (natural resins such as kauri dry slowly).

A small demand for natural resins remains, however, since they do have qualities not completely duplicated by synthetics and kauri resin has a higher drying capacity than other natural resins. This demand resulted in renewed interest in locating and extracting kauri resin from the land. In 1973 the company Kauri Deposit Surveys Limited was incorporated for the purpose of using modern methods to extract fossilized resin from the ground. They chose the Kaimaumau peat swamp near the city of Kaitaia in the Northland as the first of their extraction sites. This great swamp stands on the site of an ancient kauri forest, and remains of the trees as well as resin occur up to thirty-five feet below the surface. Although the gum diggers did exploit the swamp for large pieces of kauri copal, they dug down to only about ten feet. This new company wanted to recover the remaining pieces of gum in the swamp. They planned ultimately to extract some 10,000 tons of copal per year, having estimated that more than 500,000 tons valued at $175 million (NZ) existed in the Kaimaumau deposit.

In 1985, Kauri Deposit Surveys Limited became known as Kaurex Corporation Limited and expanded its interests to waxes as well as resins. In a later prospectus (June 1985), Kaurex estimated that by working 1,450 hectares of peat from Kaimaumau swamp over a period of about thirty years, they would recover some 600,000 tons of resins and waxes, which would be valued in excess of $1,200 million (NZ) on world markets. The resin volumes would represent 1 percent of current world resin exports, and the wax would represent 4 percent. The products would go to blenders, refiners, and formulators of resins and waxes in Europe, the United Kingdom, Japan, and the United States. Uses for the resin would include printing inks, industrial moldings, surface coatings, chemical and rubber compounding, and adhesives.

The company had access to 5,000 hectares of swampland and employed fifty workers at its peak. Problems with the extraction process hampered results, however, and in 1988 Kaurex Corporation Limited went into receivership.

Aside from resin collected directly from wounds inflicted on the trunks of the trees, all of the copal collected beneath the earth's surface during this picturesque period of New Zealand history was Pleistocene in age. The carbon dating of kauri wood, in association with the copal at various localities, provided a range of 850 to 45,000 years of age.

New Zealand also contains amber deposits. Most were formed in place (not redeposited in rock strata) and are therefore still associated with kauri wood, which now has become coal. Amber in New Zealand was first discovered by the German geologist Ferdinand von Hochstetter in 1864. He found fossil resin in coal seams near Auckland and Nelson, noting that the pieces ranged from fist to head size and varied from wine yellow to dark brown in color. Because the amber occurred in coal seams, it was referred to as a resinite and a special term, ambrite, was coined specifically for resinites (or amber) from New Zealand coal seams.

I was able to examine some of this New Zealand amber personally when I visited several coalfields in Otaga on December 20, 1988. Accompanied by Michel Poole, then a geology graduate student at Otaga University in Dunedin, New Zealand, I drove to the Roxburgh coalfield near the town bearing the same name. It was warm and sunny, and we had no trouble locating the fossilized resin, whose yellow opaque color contrasted sharply against the dark sub-bituminous coal beds. The pieces that had been exposed to the sun were partially oxidized, and, unfortunately for us fossil hunters, most of the material was opaque yellow and impossible to see within. Some lumps were the size of a person's head, but they usually fractured into many small pieces when I tried to pull them out of the coal seams. This material had been dated at 20 million years, or from the Lower Miocene. We continued to another coalfield, this time an Upper Cretaceous bituminous deposit at Kaitangata. The amber, in very small pieces and attached to the coal, was difficult to pry loose. Later studies, conducted with Joseph Lambert at Northwestern University, showed that the amber from both of these coalfields originated from kauri trees, possibly now extinct species since at least four fossil species of *Agathis* have been described from New Zealand.

Roberta and I drove with our son Greg through New Zealand, from tip to tip, examining the various coalfields and removing samples of amber, when present, for later analytical studies. We collected amber from no less than seventeen New Zealand sites, ranging from Miocene to Upper Cretaceous. In a few cases, the amber had been separated from its original coal source and had been redeposited in various types of sedimentary rocks. We found pieces of amber embedded in sandstone at Hikuranga, in claystone at Waikata, and in limestone at Raglan. Results of our analytical studies showed that all of the amber originated from kauri trees.

A most interesting point was Curio Bay at the southernmost tip of the South Island. We could stand on the beach and gaze out at an ancient forest under the sea. How strange it was to see the waves crashing over the broken stumps and fallen logs of this long-vanished forest. Wading out to a log that looked as if it had fallen into the ocean just last week, I bent down and

A milliped naturally entombed in kauri copal.

touched it. It was petrified, as they all were, and they were not from last week but from the Jurassic, some 140 million years ago! Although no one knows exactly what kind of trees they were, some refer to them as *Araucarioxylon,* based on sections of the wood, whereas others consider them the oldest known kauri pines in New Zealand.

Very few natural biological inclusions have been reported from New Zealand fossilized resin. This is strange, since tests have indicated that other types of kauri pines produced the Canadian and Lebanese amber, in which insect fossils can readily be found. Many pieces of kauri copal contain biological inclusions that were intentionally placed there by humans, and it is often impossible to be sure which are natural and which are man-made. Identification of the inclusion cannot tell us for certain, because the copal is usually less than 50,000 years old—not old enough to outdate the life span of most invertebrate species. Inclusions placed in kauri gum (or kauri copal) tend to be large and showy. A group of these on display at the Kauri and Pioneer Museum in Matakohe includes an adult huhu beetle (a long-horn beetle native to New Zealand), a monarch butterfly (an exotic species that arrived in pre-European times), snails, fern leaves, cockroaches, wetas, scarab beetles, earwigs, crane flies, and spiders. Skinks and geckos also were placed in kauri copal and often later sold as fossils in Baltic amber. (No lizards have ever been reported in real, natural Baltic amber. However, both anoles and geckos occur in Dominican amber.) The presence of non-native insects in kauri copal is generally a telltale sign that they were man-made. Honeybees, for instance, were not brought into New Zealand until 1829, and their presence (as well as that of yellow jackets) in kauri copal indicates that they are fakes.

The art of placing insects in kauri copal in a natural-appearing manner is a well-guarded secret among forgers. It is said that the old gum diggers used to heat the copal on their shovel blade over the campfire and insert objects when the resin became soft. Another method involved scraping out a cavity, placing the object to be entombed in the excavation, and filling the cavity with melted resin.

Why are natural biological inclusions like insects so rare in kauri copal? A dearth of insect life in the kauri forests could be the reason. In Baltic and Dominican amber deposits, ants and small flies make up at least a third of insect inclusions. New Zealand supports only ten species of native ants, however, and only a single native fungus gnat; this would explain the absence of these groups in kauri copal. That the resin may have formed a crust quickly, thus losing its ability to trap insects, is another possibility. In addition, much of the copal dug out of the ground is opaque, thus masking any inclusions that might be present.

Despite the above, natural invertebrate and plant inclusions do exist. But they are rare: aside from two specimens we discovered, and others in various museums and private collections, only one published report of a beetle, spider, and millipede in New Zealand copal could be located; and there are no reports of fossils in New Zealand amber.

Entering Australia after New Zealand was like moving from one prehistoric world into another. In contrast to New Zealand, Australia contains a wealth of native mammals. Yet these are mostly marsupials, and they provide evidence of how Australia itself was isolated for some 45 million years from continents containing placental mammals. Some 30 to 60 million years ago, much of Australia was a rain forest, with vegetation similar to that existing today in parts of Indonesia and New Guinea. Today what little rain forest remains in Australia is confined to small refugia containing relictual types of vegetation. The reduction of rain-forest habitat by climatic change was exacerbated by the arrival of Europeans on the continent some 200 years ago. Native rain forests were cut for lumber, and many tree species were threatened with extinction.

A traveler in Australia encounters few reminders of the past rain forests. The drying of the climate during the past 10 million years has resulted in the evolution of certain types of flowering plants, like eucalyptus, acacias, banksias, and myrtles, which we commonly associate with that continent.

Although some kauri pines in Australia thrive in botanical gardens in Melbourne and Sydney, all natural populations are confined to small areas in Queensland. These include the smooth-barked kauri (*Agathis robusta),* bull kauri (*Agathis microstachya*), and blue kauri (*Agathis atropurpurea*). These kauri pines are among the largest trees in the Australian rain forests, attaining a height of 160 feet and stem diameter of 7 feet. Magnificent trees, which unfortunately produce desirable wood for furniture and cabinet makers, they have thick columnar stems and spreading crowns.

Kauri pines have existed for a long time in Australia. Not far from Sydney, in New South Wales, a fossil site known as the Talbrager Fish Beds contains, in addition to fossil fish, an interesting array of fossil plants. Apparently, this ancient lake, Talbrager, which existed in the Jurassic period some 175 million years ago, occurred within a kauri forest, mixed with other conifers, cycadophytes, and seed, tree, and ground ferns. Leaves of all of these plants fell into the lake and then sank to the bottom, where the oxygen was low and thus the decay rate was slow. Eventually, they were covered with silt, which eventually turned to shale.

Mary White described the Talbrager fossil kauri pine as *Agathis jurassica* in 1981, and it appears to be the oldest known fossil of its kind.

Arriving in Melbourne on February 23, 1989, we first visited the Victoria Museum and discovered that amber had been collected from the Wonthaggi and Yallourn coalfields. Before leaving for these sites, we visited some of the opal shops in the city. Although the opals themselves are strikingly beautiful, with a remarkable range of colors, the opalized fossils were especially attractive to us. We marveled at the beautiful shells, teeth, bone, twigs, conifer cones, fish, reptiles, and even dinosaurs and plesiosaurs that had been wholly or partly replaced with the iridescent quartz-like gemstone, opal.

As soon as we left Melbourne and headed southeast toward Wonthaggi, we encountered a fly, about the size of our housefly, that was attracted to our faces, especially to the moisture around our lips and eyes. We soon learned that these bush flies were a common nuisance in Australia, and that the waving Australians we saw weren't being especially friendly to us but were merely brushing off the insects.

The State Coal mine at Wonthaggi closed in 1968 but is still maintained as a tourist and historic attraction. As we walked down the 180-foot tunnel to the first coal seam, I noticed that this was a hard bituminous coal, commonly known as black coal. This grade of coal normally has been subjected to high temperatures and pressures, which destroys much of the associated amber. The Wonthaggi coal dates back to the Early Cretaceous, and much of it seems to have been formed from ferns and conifers. Once in the coal seam, we examined the coal beds for resin, here known as coal gum. There was a layer of coal at the base, then an overlayer of mudstone, followed by a layer of sprint (coal subjected to a higher temperature), and finally, a layer of sedimentary rock, some layers of which contained carbonized remains of fern leaves. We could find no fossilized resin in the mine, nor in the sprint heap, and so decided to continue our trip toward the Latrobe valley, where the major deposits of brown coals occur in Australia.

These brown coals are also known as lignitic coals, because they are still at an early stage in the coal-forming process. This means that they haven't been subjected to the high temperatures and pressures that could destroy any associated fossil resin. On reaching the Morwell and Yallourn open cut mines, we were amazed at the amount of coal present. Some coal seams ranged up to 255 meters in thickness, and if you consider that a seam this thick probably took over a million years to form, you have some idea of the time involved in producing a deposit like this one.

We went down into what is called the Morwell No. 1 open cut mine, which is 70 kilometers long, 16 kilometers wide, and 600 meters deep—probably the largest open pit coal mine in the world. We stood at the bottom of this huge expanse, surrounded by towering cliffs of coal constantly being cut away for fuel for the electric generation plants. Power stations built right on these coalfields provide the bulk of Victoria's electricity supply.

Throughout the coal layers, we found variously sized pieces of amber. Material that had been exposed to the high surface temperatures, arid conditions, and bright sunlight had become highly oxidized on their outer surfaces. But material we pulled out of the coal seams, where it had been protected from the elements, was solid. Fossil leaves occurred in the mudstone separating the coal seams, but animal fossils were nonexistent. Presumably, the acid nature of the coal deposits had decomposed any animal remains.

These coal deposits, ranging from 20 to 50 million years old, apparently were derived from large forests that flourished in the region of eastern Victoria. The forests also included many kauri pines; tests by Joseph Lambert and his associates, conducted with nuclear magnetic resonance on amber from these mines, supported earlier conclusions that the material was derived from *Agathis* trees. Along with the kauri, podocarps and a mixture of angiosperms as well as ferns and liverworts grew in the forest. The coals were formed in vast lowland peat swamps, which were intermittently inundated by the sea. Tree stumps still in place show that the coal was formed in situ, and the physical factors that formed the coal were obviously the same that had fossilized the original plant resin.

We watched as the huge earth movers crushed the coal, together with the amber, and blew it into furnaces to remove the moisture (brown coal contains up to 60 percent water). The material would then be placed in machines and made into briquettes. To our knowledge, none of that amber has been systematically examined for fossils.

We never did visit the natural stands of kauri trees in Queensland, but from what we learned, kauri copal was never mined in Australia and we couldn't find any record of its use by the indigenous peoples.

16 Rekindling the Quest for Ancient DNA

After Svante Pääbo left Berkeley for Munich, and Allan Wilson became ill, the ancient DNA studies came to a halt. All our previous work seemed to amount to nothing, destined to become simply a footnote in the annals of science. Fortunately, however, new developments rekindled our optimism.

In the fall of 1991, our son Hendrik, just beginning his senior undergraduate year at California Polytechnic State University, San Luis Obispo, enrolled in a molecular biology laboratory. He came up to Berkeley for a weekend in September, and during our visit he asked me what was happening with our DNA amber project. I explained that it had become essentially dormant. Hendrik excitedly replied that he would like to carry on the project, and that his professor in molecular biology, Dr. Raul Cano, would help him with the initial extraction process and supply the laboratory facilities. I agreed that it would be worth a try, and a few hours later Hendrik left in his 1966 Volkswagen bug headed back to Cal Poly, carrying five stingless bees in Dominican amber.

It didn't take him long to trim the amber pieces so that they would lie flat in the bottom of a petri dish. But how should he reach the tissue remains inside the bee bodies? Hendrik reasoned that there must be a better way than grinding the entire piece up and placing everything in extracting solution, as we had already done.

Now and then something else you are doing, completely unrelated to the project at hand, sparks a new idea. Hendrik also happened to be working on a project involving the examination of liquid on the surface of human lung tissue. When frozen in liquid nitrogen, the tissue formed breaks

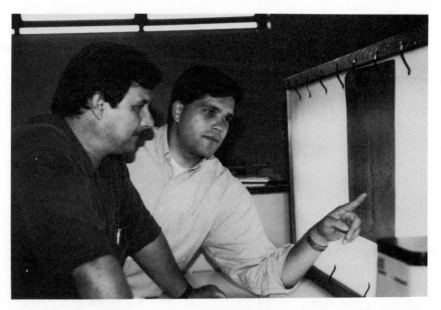

Raul Cano (left) and Hendrik Poinar examine the first DNA sequences from an organism (stingless bee) in amber.

A stingless bee in amber that was split open using the liquid nitrogen treatment developed by Hendrik Poinar.
(Photograph by Hendrik Poinar)

A stingless bee in Dominican amber.

along the fracture planes of the lungs. And it suddenly occurred to Hendrik to try this same procedure with amber.

When amber, a poor conductor of heat, is covered with liquid nitrogen, the surface layers cool more rapidly than the inside. This results in various fracture planes running through the piece. By some coincidence, very faint pressure planes are formed by insects embedded in the amber. These pressure planes become fracture planes when the amber is supercooled by liquid nitrogen. When warm saline is then added, the sudden reversal in temperature causes the fracture planes to break apart. In many cases this results in the amber cracking right through the insect inclusion, exposing the tissue needed for DNA extraction.

Once this process had been perfected, it allowed Hendrik to crack open the insects, scrape out the mummified internal tissue, and transfer it immediately to DNA extraction solution. This important breakthrough allowed us to obtain large amounts of internal tissues in a relatively short time.

The polymerase chain reaction, mentioned previously, also allowed us to shorten the time period in working with ancient DNA. This reaction is like a genetic photocopying machine that copies a target gene introduced into the system. Thus small amounts of ancient DNA extracted from amber inclusions are the target genes, and after two hours of amplification by the polymerase chain reaction, enough copies (millions of them) are made to proceed to the next step of sequencing. The latter involves breaking down the DNA strands into their respective bases and obtaining the genetic code. Hendrik and Raul Cano were lucky: they obtained DNA from the very first stingless bee they cracked open. Hybridization studies conducted in November 1991 showed that the DNA was not human or bacterial, the two most likely sources of contamination. These initial results were published in April 1992. In the meantime, the DNA strands were being sequenced and compared to homologous strands of living stingless bees in Panama that were furnished to us by David Roubik at the Smithsonian Tropical Research Center. The sequencing studies were begun in December 1991 and finished in March 1992. In all, we examined five fossil bees. The results showed that DNA strands extracted from the fossil bees most closely resembled similar strands of DNA from the Panamanian bees. We now believed that we really had obtained DNA from a stingless bee (*Proplebeia dominicana*) that had been preserved in Dominican amber for 25 to 40 million years. These new results were published on September 1, 1992. Both of these papers attracted much attention, because they represented the oldest known DNA at that time. Shortly after our last paper, Rob DeSalle and his colleagues at the American Museum of Natural History published their succcessful extraction and sequencing of DNA from an extinct *Mastotermes* termite in Dominican amber.

This initial success raised the question of how long DNA could be preserved inside insects in amber. If we could obtain some older amber from the Cretaceous period, it might be worth looking for DNA in some of those insects, even though the amber was produced by trees other than those responsible for Dominican amber.

Back in 1982, Aftim Acra, a professor of environmental sciences at the American University of Beirut in Lebanon, had sent me some pieces of Lebanese amber for scientific study. Lebanese amber is especially interesting because of its antiquity, dating back to the early Cretaceous, some 120 to 135 million years ago. In fact, it is the oldest amber known to contain insects. One piece of Lebanese amber that Acra had sent contained a weevil.

The successful results of extracting DNA from this insect are reported in detail in the next chapter.

After recovering insect DNA from amber inclusions originating from two different geological periods and geographical areas, the question arose as to whether plant DNA could also be preserved in amber. At that time, the oldest reported plant DNA came from leaves embedded in layers of rock from an ancient forest that existed some 17 to 20 million years ago. These deposits, called the Clarkia beds, are located in Idaho and contain a range of fossilized leaves.

In March 1993 I sent Hendrik a piece of Dominican amber containing a leaflet of *Hymenaea protera,* the extinct tree responsible for much of the amber in the Dominican Republic. The plant material posed some unforeseen problems. In contrast to insect DNA, nucleic acids from plants are often associated with powerful inhibitors that bind to the DNA and prevent amplification. Hendrik soon learned that various compounds could be added that blocked the action of these inhibitors and allowed the DNA to be extracted and amplified for sequence studies. Using these methods, Hendrik sequenced DNA from the ancient *Hymenaea* in Dominican amber in April 1992. In May I sent him samples of recent representatives of this group from living and herbarium specimens. These were sequenced in October 1992 and compared with those of the fossil plant.

Our hopes that fossil DNA was preserved in plants in amber were realized. In addition to discovering the oldest known plant DNA, the molecular studies supported earlier surprising conclusions based on traditional morphological studies, namely that the Dominican Republic amber plant was more closely related to present-day populations in East Africa than to similar plants in South America.

Teamwork in scientific research not only is advantageous but in many cases is crucial. A group of well-qualified individuals working together can almost always accomplish much more than individuals working alone. We were fortunate to be able to assemble a team whose combined expertise included a knowledge of amber fossils, DNA extracting techniques, reading sequences, sequence analysis, phylogenetic reconstruction, and collecting recent species closely related to amber fossils. We owe our success to this combined effort.

17   Going Back in Time

The oldest DNA recovered to date came from the weevil encased in Lebanese amber originally sent to me by Aftim Acra on May 15, 1982. From the early Cretaceous period, this piece was one of several in the Acra collection that I cursorily examined and then put aside because of other pressing studies. Not until ten years later, after we had already extracted the first DNA from an amber insect, did I remember the Lebanese samples and turn my attention to those precious pieces of fossilized resin. Perhaps we could extract DNA from insects in 125-million-year-old amber. But I couldn't find any complete insects in the amber I had—only insect fragments. We needed a complete insect with internal tissue in order to proceed.

One piece of amber was a bit larger than the rest, however, about a centimeter in diameter. It was composed of concentric layers of resin, one deposited on top of the other, and although I couldn't see a specimen within, I slowly began to remove the layers, beginning with the outer, one after the other. After the third layer, I saw a black object between the fourth and fifth layers. I increased the magnification of the microscope. I could make out the outline—it was a beetle. I turned the piece slightly and viewed it from another angle—it had a snout. It was a weevil!

When I realized that no weevils had ever been described from Cretaceous amber, I concentrated on obtaining a clear view of the insect in order to further identify it. As I carefully lifted up the piece of amber, the fourth layer broke away from the remainder of the piece, splitting the weevil in half. My heart sank. This beautiful weevil was broken. Then, to my astonishment, I noticed that a pellet of tissue, preserved for so long a time, had

Lebanese amber weevil from which the oldest known DNA was extracted.

slipped out of the weevil's abdomen and lay on the amber. I couldn't believe my eyes. The weevil was tiny (only three millimeters long), and the pellet of tissue was only half that size. I wracked my brain to decide whether I should preserve it dry or in 100 percent ethyl alcohol. After a discussion with Roberta, we decided the alcohol would be better; at least the pellet would be easier to retrieve. So I placed the small pellet, which later turned out to be largely ovarian tissue, in a small vial of 200-proof ethanol, adding the two amber pieces containing the remainder of the weevil (I could see that some additional tissue still remained inside the insect's body). I called Hendrik and told him to prepare to receive the specimen, which I sent to him via Federal Express the next day.

Hendrik received the package on June 24, 1992, opened it, and carefully removed the small pellet. First he photographed it; then he removed it from the ethanol and placed it directly into a chelex (dissolving) solution to extract any DNA that could be revived after 125 million years. After tossing and turning the tissue in the extracting solution for twenty-four hours, Hendrik removed the liquid solution and immediately began the amplification process using the polymerase chain reaction. An excited telephone call from him on June 26 informed me that there was DNA in the sample, which was good, but we had no idea where the DNA had originated. During the next week, Hendrik worked with a variety of primers with different concentrations of the co-factor magnesium chloride, in order to obtain sufficient amounts of DNA for sequencing.

The sequences were completed on July 4, 1992, and we anxiously waited while the computer scanned the DNA fragment and made a comparison with corresponding sequences of modern-day organisms from GenBank®. Would it turn out to be a human contaminant? The computer showed us that the closest known DNA to our fossil belonged to a beetle in the same order as our weevil. This meant that we had more work to do, for the beetles comprise a large group: There are ground beetles, click beetles, leaf beetles, ladybird beetles, bark beetles, and many others. The weevils themselves comprise several families. We had to narrow it down to the weevil group (Curculionoidea) and, we hoped, to the same family as the fossil (Nemonychidae). Somehow we had to obtain DNA from the closest living descendant of the extinct weevil as possible.

This meant that we had to repeat all the extractions, amplifications, and sequencing that we had done with the fossil weevil, only now with living beetles. Weevils are common enough insects, and, fortunately, members of the Entomology Department at Berkeley were rearing one notorious member of this group, the alfalfa weevil, which has destroyed tons of alfalfa plants over the past twenty years in North America. So I called Lou Etzel, our quarantine expert, and obtained some alfalfa weevils from him. Could we find something still closer to the fossil weevil belonging to the family Nemonychidae? Its members even today are tiny and difficult to find. Most develop in the male cones of conifers where the larvae feed on pollen. Where could we find some?

At the insect museum at Berkeley, with the help of our beetle expert, John Doyen, we located some dried, pinned specimens of a small nemonychid weevil known as *Lecontellus* that had been collected from pine trees in California. Could we obtain DNA from these dried specimens that had been collected some twenty years ago and surrounded by preservatives ever since?

As soon as we had finished our DNA studies with the fossil weevil and could no longer contaminate the samples, I sent living specimens of the alfalfa weevil and dried specimens of *Lecontellus* down to San Luis Obispo. These were received on July 5, 1992. Everyone was so eager to learn the results that Raul Cano and Hendrik worked around the clock, extracting, amplifying, and sequencing DNA from both the living and dried weevils for the next two weeks. The sequences were entered into the computer and run against all of the known animal sequences. We waited as the screen flashed base pairs back and forth—then silence. We looked. The extinct DNA was closest to the extant member of the Nemonchidae family, *Lecontellus.* The alfalfa weevil was next closest. We could hardly believe our eyes. What had begun as a long shot had materialized. We had awakened fragments of insect DNA that had remained dormant for over 100 million years!

Meanwhile, the remains of the fossil weevil had been sent to the weevil expert G. Kuschel in Auckland, New Zealand. Fortunately, even though the specimen was broken into two parts (and later, as Kuschel was examining it, one of these broke again), enough characters remained to determine to what family and subfamily this creature belonged. Kuschel found that it was a female and belonged to an extinct subfamily (Eobelinae) previously known only from the Jurassic era. Since the fossil could not be placed in any of the known extinct genera, Kuschel erected the new genus *Libanorhinus* and new species *succinus* for this fossil. The genus name is derived from the Lebanese mountains that the Greeks called *Libanos* and the Romans called *Libanus,* and from the Greek noun *rhinos,* meaning nose or weevil. The specific name *succinus* was derived from the Latin noun *succinum,* meaning amber. After the descriptive study was finished, I glued the three amber pieces containing the weevil back in place and embedded the composite piece in clear plastic for safekeeping.

How did *Libanorhinus succinus* end up in a piece of amber? That scenario can be reconstructed from what we know about the modern descendants of nemonychid weevils. One of the foremost behavorial paleontologists, Art Boucot from Oregon State University, spent years collecting literature on the evolutionary behavior of extinct organisms. He came to the intriguing conclusion that fossil organisms closely related, morphologically, to present-day descendants probably exhibited types of behavior similar to their descendants. In other words, there is a certain fixity in general behavior patterns. Following this line of reasoning, if we examine the behavior of present-day nemonychid weevils, we discover that of the twenty-two known genera, all but two develop on pollen in the male cones of conifers. Since conifers were the predominant higher type of vegetation in the early

Cretaceous, when Lebanese amber was being formed, one could imagine that nemonychid weevils were fairly common. In fact, the trees that produced the resin that became Lebanese amber were themselves conifers of the family *Araucariaceae.* So it is highly probable that the female fossil *Libanorhinus* was searching for male cones of a kauri pine on which to lay her eggs when she landed on a pool of resin on the bark or a branch of the tree. It was unfortunate for the weevil, but quite fortunate for us, some 125 million years later.

The first historic record of Lebanese amber appears in the writings of the German biologist Oscar Fraas, who in 1878 discovered that this material dated from the Cretaceous and had insect and plant remains trapped inside. This amber was "rediscovered" in 1962 by a group of scientists looking for fish fossils in the Bekáa valley, near the town of Dar el Beida. The rediscovery has led to two recent major collections of Lebanese amber— the Acra collection in the United States and the Milki collection maintained at the American University of Beirut.

On July 23, 1993, I met in New York with Aftim Acra, who was a member of the group that made that important 1962 trip to Dar el Beida. He related how he had acquired his collection over the years. The area where the amber is deposited, between Lebanon and Israel, is a desolate one nearly barren of animals. Everything but a few mountain goats has been killed off by the hunters or soldiers who pass regularly through the area. The vegetation is equally sparse. Some small, scraggly pine trees sit high on some out-of-the-way bluffs, but most plant life consists of shrubs, grasses, and a few wildflowers that grow around the scanty springs trickling out of the hills. The area offers a stark contrast to the original flat floodplain of 125 million years ago, which was lush and semitropical, filled with plant and animal life. At that time, large kauri pine–like trees flourished under the moist climate and dinosaurs dominated the land.

Humans inhabited these lands some 100,000 years ago, long after the mountains had formed and the climate had become dry and arid. While searching for other amber sites, Aftim came across human remains dating back 30,000 years, and discovered Phoenician bracelets of copper and bronze, and even an entire elegant blue glass vase made by these famous craftsmen and traders. At several locales Aftim found the remains of coins, lamps, and bracelets that the Romans had abandoned in the Lebanese hills. Byzantine crosses and other artifacts identified other cultures that had existed, persisted, and perished in the now bleak domain.

Aftim developed an eye for archaeological remains as well as amber. One day he discovered what might be called the oldest garbage dump in the world. Observing British archaeologists excavating a cave in the vicinity of the amber sites, he watched as they carefully sifted through several feet of soil that had composed the original cave surface. After they finished their dig, crated their findings, and left for London, Aftim discovered an area outside the cave entrance that had served as a waste disposal. And in this pile of refuse lay treasures much more extensive than what had been recovered from the cave floor. Between the layers that had formed since the last inhabitants had resided there some 30,000 years ago were arrowheads, pottery fragments, and the bones and teeth of animals that had served as food for these peoples. Perhaps these past inhabitants had observed the amber buried in the hillside; and maybe they had discovered and marveled at the tiny life forms enclosed inside. The fragility and brittleness of the pieces may have discouraged the local artisans from attempting to form them into amulets or beads, as was done along the Baltic coast.

Aftim and others went frequently into the hills of southern Lebanon looking for amber. In the evening and on weekends, when he was not involved in teaching or administrative duties at the American University of Beirut, Aftim would wash his collected treasure in water, dry it with a towel, and then patiently polish the pieces. This was an exceedingly slow, tedious process, since much of the amber was highly fractured from pressure exerted on the fossilized material during the millions of years it was slowly moved within the earth. Thus many of the pieces broke, even when handled ever so gently. But Aftim continued, rubbing them across pieces of wetted sandpaper on the surface of the table and then polishing the surface on a piece of terrycloth. At first Aftim entrusted a group of pieces to an expert jeweler in Beirut. When, after a few weeks, most of the pieces were returned broken and still unpolished, he decided to handle the process himself. Eventually, he found additional amber sites, discovering that some areas were much more prolific than others. Amber so fragile that it started to crumble was mounted in Canada balsam. Slowly the collection grew; and when the press heard of it, they mistakenly announced that Aftim was making a "storage place," which the Arabic word for amber (*anbar*) also means.

Then came the civil war. Fieldwork during that time could be dangerous and had to be planned carefully, since unexpected contact with soldiers could occur anytime. This conflict, like many, pitted brother against brother, family against family, and the number of factions was so great that no one could be sure who the enemy was. Children who had played together in the street a few years before now had guns and took charge of the neighborhood activities. Digging anywhere was considered suspicious—one might be planting land mines or looking for them. Nobody would believe that anyone would waste his or her time looking for tiny fossils in pieces of fossilized resin, especially when a war was raging. In fact, because of land mines, it became too dangerous to search for amber, and Aftim limited his activities to sorting, shaping, and polishing his precious pieces.

Nowhere, even in Beirut, could a person feel safe. The windows of the Acra house were constantly shattered by bombs. Finally, one day, Aftim and his wife couldn't stand it anymore. They secured a car and drove off to Damascus for some peace. They returned from this short reprieve only to discover that their home had been broken into several times and some amber had been removed, along with other precious archaeological remains. Most of the collection was still there, however, and Aftim realized that he should photograph each fossil as documentation in case they were ever stolen or destroyed. Even his laboratory at the university was not safe. Thieves broke into the rooms and took all of the scientific instruments, including microscopes. In his haste to finish, he was still photographing amber on the day the Israeli army entered Beirut.

Part of the Acra collection was brought to the United States in 1976 by Fadi Acra, Aftim's son, and the remainder arrived in 1982. It was in the latter year that Aftim became aware of our studies on amber insects and wrote to us with the hope of entering into collaborative studies. Curiously enough I had written a letter expressing the same wish to Aftim one day

▶

The Acras examining their Lebanese amber collection.

after he wrote his. Our letters crossed in the mail, and Aftim sent the package with the weevil-containing samples to me on May 15, 1982.

On August 1, 1993, Dr. Raif Milki visited our laboratory, bringing with him a portion of his collection of Lebanese amber. We were interested in the problem of preservation of this valuable, yet highly fragile, material. Our first task was to see if embedding the small angular amber pieces in a modern-day, clear plastic would be feasible. What effect would the plastic have on the amber over time? The first results of our investigation were highly rewarding: the plastic not only protected the amber but, by entering the larger cracks, made the fossils much more visible. Although more time is needed to determine the final effect, we decided that this method had great potential for maintaining and even restoring rare amber fossils.

Dr. Milki is still actively searching for new deposits of Lebanese amber. After twenty-five years of experience, he can now tell where amber occurs by the color of the soil. These areas, all of which are over 1,000 meters in elevation, produce three general types of amber—a yellow transparent, an orange transparent, and a red opaque. Among all these types combined, 1 fossil occurs in about 500 pieces of amber. However, Milki emphasizes that the yellow amber contains more insects than the other types, with about 1 in 25 usually containing a fossil. The amber often is just lying on the surface of the ground. If you can find a vein, you can dig down and sift through a half meter of soil to find other pieces. The largest piece Milki has seen is a block of yellow transparent amber weighing 1,800 grams. Milki's current discoveries of amber deposits in the Lebanese mountains indicate that we may continue to harvest this oldest insect-bearing amber for years to come.

18   *Jurassic Park* Repercussions

One day in July 1990 I received a telephone call from an employee at Universal Studios in Los Angeles. "We are making a film of *Jurassic Park,*" the voice explained, "and want to ask you some questions about your amber research." This was the first time I had ever heard of *Jurassic Park.* I didn't even know Michael Crichton had written such a work, and so, of course, the fact that the movie rights had been sold before the book appeared was also news to me. "You, your wife, and the Extinct DNA Study Group are acknowledged in the back of the book," the voice continued, surprising me even more. What was the book about? Little did I know that this film would have a major impact on the amber world.

According to Don Shay and Jody Duncan, in *The Making of Jurassic Park,* Michael Crichton had written an original screenplay in 1981 about the creation of a genetically engineered dinosaur. He put the manuscript aside for several years, when a different approach to the idea finally presented itself suggested by, among other things, our 1982 *Science* paper. But, due to the dinosaur mania that was sweeping the country at the time, he decided the subject would be too trendy. Then, in 1989, circumstances led him to return to the abandoned screenplay and develop it into a novel, using the theme park concept. He wrote several drafts from the viewpoint of different characters and sent the final manuscript to the publishers in May 1990. A few days later, Hollywood expressed interest, and less than a week after it was offered for sale, Steven Spielberg and Universal had accepted it.

As soon as the book became available, I read it quickly. Journalists wanting to know if dinosaurs could really be cloned were already calling me. Others wanted to know how it might be done if it wasn't now possible.

A mosquito in amber.

Amber and genetic engineering were suddenly in the news, and in a big way. Our work continued as preparations for the film *Jurassic Park* were under way. And by coincidence, the British journal *Nature* published our latest results—on obtaining the oldest known DNA from an insect that had lived during the dinosaur period—on the day before the film *Jurassic Park* was first released in North America. The news about the recovery of DNA from a Cretaceous weevil gained instant popularity. In fact, according to the Associated Press, this story made the front pages of 257 newspapers in the United States and 400 newspapers worldwide. According to them, it was the most complete coverage of a single science news item in the past twenty years.

Can a dinosaur be cloned?

Thus far we have recovered only small pieces of DNA from insects and a plant in amber. However, recovery of DNA from a Lower Cretaceous weevil that existed along with the dinosaurs shows that recovery of some dinosaur DNA—either from slightly fossilized dinosaur bones

Dr. Raif Milki (left) and George Poinar examine a blood-sucking fly in Cretaceous Lebanese amber that has been embedded in plastic for safe keeping.

or from dinosaur blood cells in amber-preserved insects—may well be possible.

Even small snippets of DNA from larger animals such as dinosaurs could help us to solve some controversial questions in the news today. To what present-day group are dinosaurs most closely related? Were dinosaurs warm blooded? What caused their extinction?

Would it be possible to obtain dinosaur DNA from bloodsucking insects in amber? Theoretically, yes; practically . . . maybe. Deposits of amber from the dinosaur period do exist (we have already discussed those from Lebanon and Canada), and they do contain bloodsucking insects. Biting midges encased in amber from the time of the dinosaurs have been known for at least fifty-seven years. They were described in Canadian Cretaceous amber by M. Boesel in 1937. D. Schlee and H. Dietrich reported a biting midge in early Cretaceous Lebanese amber in 1970 and V. Zherichin and I. Sukacheva discussed biting midges in Cretaceous Siberian amber in 1973. Such insects could have fed on dinosaurs. Obtaining dinosaur DNA from

Specimens similar to this blood-sucking phlebotomine sand fly in Dominican amber also occur in Cretaceous Lebanese amber and may have fed on dinosaurs.

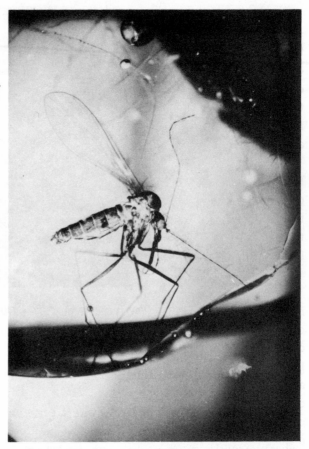

amber insects is confounded by the problem of verification. How do we determine whether any vertebrate DNA recovered is really dinosaur DNA when there is nothing around for comparison? A much more satisfying scenario, which may or may not be feasible, would be first to recover DNA from known dinosaur remains (bone, skin, teeth), and then to examine bloodsucking Cretaceous insects.

Extinct multicellular life forms such as dinosaurs cannot be brought back to life today for two major reasons. First, all DNA recovered from fossils, including those in amber, is damaged. If the DNA is incomplete, important bits of information needed to make a species are absent. There are

DNA repair systems, but until we know exactly what is missing and what can be repaired, nothing can be done. And even if we had the complete DNA of an extinct animal, in order to initiate development and make another individual we would have to make the DNA feel that it was in an embryonic state when we inserted it into a host egg. Then we would have to make the egg develop normally without transforming it into an undifferentiated mass of tissue. Thus far, biologists have been unable to clone an individual from a specialized body cell.

When, and if, we become able to re-create extinct species, we will be faced with a moral dilemma. Is it ethical to bring back an extinct species out of scientific curiosity? The re-created species may have been adapted to a past environment completely foreign to those now existing. If a plant feeder, it may have existed on vegetation now extinct. Even with the best DNA repair system available, there is a good chance that not all physical and mental attributes would be completely normal.

And would extinct life forms brought back to life pose a danger to modern-day organisms? The cases today of humans moving animals from one continent to another in an effort to do good, but with unfortunate consequences, are many. Consider the introduction of the mongoose into Hawaii for rat control: the mongooses' appetite went far beyond that of rats and reduced some native bird populations to the point of extinction. The cane toad was introduced into Australia to eat beetles that attacked sugarcane in the northern provinces—but this huge amphibian multiplied, soon outran the canefields, and ate not only beetles and other native insects but endemic amphibians, lizards, and birds. The gypsy moth that defoliates acres of trees in eastern North America was imported by unsuspecting scientists in order to provide an alternative source of natural silk. The proposed gypsy moth silk industry never materialized, however, and the damage and control of the escaped moths led to considerable expense in terms of both time and money. The effects of the accidental introduction of the African honeybee are now being experienced in the southwestern United States. Wanted initially because it actively produces more honey, this bee possesses a highly aggressive nature that has led to attacks on wildlife, more docile bees, and even humans.

The rest of this book could be filled with additional cases of humans, either intentionally or accidentally, introducing animals or plants from one area into another with harmful results. Individuals and programs must be responsible for their deeds. We can agree now that the above, and many other introductions, were wrong or bad, but hindsight does not change the consequences, nor can it improve the ability of humans to predict the future effects of their actions. When we consider reviving extinct life forms, let us not forget these past scenarios.

19 Amber from the Dinosaur Period

A cloud of large, metallic-eyed horseflies circled around my head as I clambered over the small outcrops of Late Cretaceous coal deposits in southern Alberta, Canada. This had been dinosaur country 70 to 80 million years ago, and the evidence now, in July 1993, was plain enough. Here and there, scattered among the pieces of coal, lay bone and tendon fragments of large duck-billed dinosaurs that had fed on the vegetation comprising the subtropical-tropical forest that existed back then. I stood up and looked at the area now, wondering when the dark, rain-filled clouds overhead would begin to release their moisture. The tall silken stems of spear grass moved en masse like ocean waves under the winds that caressed the rounded hills. Their soft whisper was the music of this prairie. Here and there flowering shrubs of snowberry showed their round white blooms. In the valleys between the low hills were ponds cradled by reeds and lush grass holding nests of blackbirds. Clumps of cottonwood trees kept their distance, but not too far, from the only water for miles around.

The pungent smell of sage rose on the wind. These plants provided a resting place for the pygmy grasshoppers that were being searched out by the bold magpies. A rabbit sat at the entrance of its den among a pile of moss- and lichen-covered gray and brown stones. Its coloring made it appear as yet another stone; only the moving ears revealed its identity. But it was gone in a flash when a hawk that was circling on the wind back and forth let out a loud screech.

This unique scene was quite different from what it must have been 75 million years ago. I tried to imagine large sequoia and *Araucaria*-like conifers where the cottonwoods now grew, and palms and cycads lining the

The authors holding Canadian amber dating from the Cretaceous period.

Greg Poinar collecting Canadian amber in Alberta.

edge of the water where reeds and cattails now existed. Back then there were giant ferns and horsetails, ginkos and flowering plants, that had long since disappeared. There were no rabbits at that time; at least, no remains have been found. From the bones, shell fragments, teeth, and footprints, we know that the area was probably dominated by dinosaurs, but near the water's edge they probably fought for space with *Champsosaurus,* a large alligatorlike reptile, and soft-shelled turtles.

My vision of antiquity suddenly vanished as I felt the painful bite of a horsefly on my neck. The tropical forest had long ago disappeared, together with its inhabitants; one could lament its passing but never bring it back. Amid its ruins at my feet lay its tears—amber. The rain and wind, sisters of erosion, had slowly removed the lighter coal and shale particles on the sides of the hills and the amber lay exposed to the sun and wind. In time it would crumble and turn to dust, along with its treasures of insect and plant life from the past world of the dinosaurs. But now, it just lay there, some pieces sparkling in the sun, others obscured by a dark mottled crust, camouflaged against the gray background.

Roberta, Greg, and I had been taken to this area through the courtesy of officials at the Tyrrell Museum of Paleontology at Drumheller in Alberta and the University of Calgary. Our companion and fellow amber researcher was Ted Pike, who was writing his Ph.D. thesis on various aspects of Canadian Cretaceous amber and had collected amber from seven sites in Canada. The deposit he took us to was by far the most prolific in abundance of amber and was also the focus of his thesis.

The actual collecting of amber pieces, most of which range from very tiny to the size of a quarter, is tedious. You walk, partly bent over, staring at the ground and constantly scanning a two-foot area in front of you. Some pieces are thick and flat, and were probably formed within the bark of the amber tree, so they rarely contain insects. Others are cylindrical or tear-shaped and represent the remains of resin icicles that flowed down the outer surface of the bark; these are the ones that contain a variety of small life forms, such as mites, aphids, wasps, ants, cockroaches, and, even more important, bloodsucking insects. Ted Pike surprised the scientific world when he discovered the oldest known mosquito from these deposits—essentially doubling the previously known fossil age of these bloodsuckers.

But the most common biters in this amber were the biting midges, or cer- atopogonids, as they are known to entomologists. Right here we could imagine our own *Jurassic Park.*

Spatters of rain suddenly pelted the ground around us, darkening the already dark sod and disappearing into the prairie soil. We zipped up our raincoats and continued, walking from one exposed hill through clumps of prickly rosebushes to another one, always looking down. In some areas, the smartweed had grown up the sides of the mounds, forming a tangle of stems that concealed the amber pieces. We came across a sandstone boul- der that carried the three-toed imprint of a large duck-billed dinosaur. Ero- sion had cut steeply through some areas, forming treacherous cliffs that would give way if we walked near the edge. Apparently a calf had done that just before we arrived and now lay twisted, its neck broken, in a tangle of rosebushes. Its mother was still standing on the cliff edge, calling for it to return to her. In the more eroded areas, the ground was honeycombed from the activity of large red ants. These active creatures did not hesitate to remind us, by crawling up and biting our legs, that we were intruding on their territory. The excitement and awe we felt when we discovered am- ber right next to dinosaur bones, champsosaur vertebrae, and soft-shelled turtle remains would be difficult to convey. We could only imagine what secrets lay within those amber pieces.

This area of Cretaceous amber is not far from Dinosaur Provincial Park (the Tyrrell Museum of Paleontology Field Station), which first opened in 1955 and can boast the largest concentration of dinosaur remains (about 300 in all) found anywhere in the world, dating from some 75 million years ago. Remains of dinosaurs in this area included the bird-mimic dinosaurs (Ornithomimidae); the foot-lizard dinosaurs (Elmisauridae); the egg-thiev- ing dinosaurs (Oviraptoridae); large tyrannosaurids such as Albertosaurus, the duck-billed dinosaurs (Hadrosauridae); the bone-headed dinosaurs (Pa- chycephalosauridae); the armoured dinosaurs (Ankylosauridae and Nodo- sauridae); and the horned dinosaurs (Ceratopsidae). One bed being dug up represented a herd of some 300 Centrosaurus, or horned, dinosaurs that had died while crossing a river. Amber has also been found in Dinosaur Provincial Park itself, although it has not been collected in great quanti- ties and has not yet been studied for inclusions. Some amber has been recovered from hills composed of dolomite, a fine clay deposit that has

Beautiful, but treacherous when wet, dolomite hills in Alberta, Canada.

In Alberta, dinosaur bone and small pieces of amber from the same era occur together.

special lubricant properties when wet. I personally experienced these prop-erties when I began to climb up a dolomite hill just after a rainstorm. About halfway up, I stopped to adjust my backpack. Within seconds I was coasting back down the hill as if I were on ice, and at about the same speed. Fortu-nately there was a nearby bush to grab, since there was no way to stop and I was headed for a group of very hard boulders.

Alberta was not the first Canadian province to report the presence of amber. In 1890, J. B. Tyrrell, employed by the Geological Survey of Canada, was shown a piece of amber by an Indian living on the nearby Chema-hawin Indian Reservation in Manitoba. Tyrrell and others traced the amber back to the shores of Cedar Lake and mentioned it in an 1890 report. In 1891, B. J. Harrington officially named the amber chemawinite, after the local Indian tribe. In 1897, R. Klebs, a German researcher of Baltic amber, investigated some of the Cedar Lake amber and, apparently not knowing of Harrington's report, named the material cedarite. Neither name ever be-came established. Most people simply refer to all amber from Canada as Canadian amber.

The Cedar Lake material represented an accumulation of years and years of amber being washed up on the shore by wave action. The amber area extended nearly two kilometers along the lakeshore and to a depth of almost a meter. The early estimates that the area contained more than 6,700 metric tons of amber prompted the Hudson Bay Company to com-mercially exploit the material. They devised a wind machine used to blow away the lighter wood debris and leave the heavier amber. During the com-mercial operation, from 1895 to 1937, more than a ton of amber was col-lected, most of which was sold to varnish manufacturers—an unfortunate loss to paleontologists. During this period a few people began to wonder where the amber had come from. There were no amber-bearing beds re-motely near Cedar Lake; the closest were in the western provinces of Sas-katchewan and Alberta. Was it possible that over the ages, small pieces of amber were loosened by erosion from beds of lignite and shale, washed into rivers, and carried by the currents over 1,000 miles and deposited on the banks of Cedar Lake? This scenario could be true, since no other plau-sible explanation has yet been offered. It is possible that amber-bearing beds occur much closer to Cedar Lake but simply have not been discovered. On the other hand, some of the small lakes that the easterly flowing Saskatchewan River passes through may also have unknown amber deposits on their shores. Canadian amber deposits could be much more extensive than they now appear.

The first scientific collecting expedition to Cedar Lake was organized by F. M. Carpenter from Harvard University in 1938. Together with C. T. Brues, C. T. Parson and two others, he collected about 400 pounds of Cedar Lake amber within three months. They devised a flotation method to remove the amber from other debris along the shore. Handfuls of shore material placed in containers of salt water left the amber floating while the heavier material sank. Transferral of the floating material back to fresh water resulted in the amber sinking while the organic matter remained on the surface.

Not every piece of amber contains an insect. A report on Canadian amber, prepared by J. F. McAlpine and J. E. H. Martin, entomologists from the Canadian Department of Agriculture, estimated that one could find an arthropod in 1 out of 50 to 55 pieces with selective collecting. With nonselective collecting of all the amber, only 1 out of 1,000 pieces could be expected to contain an inclusion. In 1940 two young Americans, William M. Legg and Hambleton Symington, spent the summer collecting Canadian amber, which later became the subject of the former's senior thesis for the Department of Biology at Princeton University in 1942. In his work, Legg tells of his trip to Cedar Lake from The Pas, then a small town with mostly frame buildings and dirt streets, a gathering place for the local Indians and a supply depot for prospectors. Paddling down the Saskatchewan River in a large lake canoe with his French-Canadian guide, Legg found the landscape rather monotonous during the ninety-mile trip. Tall reeds and willows grew along the banks, and beyond them were miles of marshland with reeds and sphagnum moss, dotted with islands of poplar. He commented on the abundance of teal, mallards, baldpates, and black terns. They spent the first night in what was called the Post, a whitewashed plank warehouse near the water with an old store and small bungalow.

After reaching their destination, Legg reported that the "amber beach" was composed of woody debris of a variety of sizes, most of which was rubbed smooth from the wave action. He recounted that the search for amber among the debris was tedious and they tried to devise methods of separating it out. After testing various methods, including a flotation box,

A bloodsucking biting midge in Canadian amber. This specimen could have fed on dinosaurs, probably on the softer skin around the nostrils, eyes, and ears or on newly hatched individuals.

they decided the best was simply to hand-gather the pieces. Legg identified representatives of nine orders of insects in the amber they collected that summer in 1940. Especially notable was the first, at that time, fossil ciliate protozoan (*Paramecium*) and the first fossil tardigrade. The latter specimen was described by Kenneth Cooper, then Legg's major professor, in 1964 as *Beorn leggi,* in memory of his student who perished in a hunting accident in 1953.

▶

Burgess Shale fossils dating from 550 million years ago, long before there were terrestrial plants that could produce resin.

Scientific studies on Canadian amber are being continued by Ted Pike of the University of Calgary. One of his special finds was parasitic mites attached to the backs of biting midges, showing that Jonathan Swift's poetic line about a flea that "hath smaller fleas that on him prey" applies to the past world as well.

Although western Canada has long been well known for dinosaur fossils, a locality in Yoho National Park in British Columbia recently has become famous as another fossil site, dating from a much earlier period and containing strange invertebrate life forms. The fossils thus far discovered at this site, commonly known as the Burgess Shale, represent the greatest variety of soft-bodied organisms from a Cambrian Sea ever known. Amber is dependent on a plant source that produces the original resin. The oldest amber we have found with recognizable fossil inclusions comes from the Lower Triassic, some 225 million years ago. The oldest known amber has been reported from Carboniferous coal seams, some 300 million years old, but this amber is difficult to obtain. Since land plants first appeared in the Devonian, 380 million years ago, some of them might have produced resin that persisted up to the present, but there are no reports of amber from this period. No fossil record exists of land plants from the time when the Burgess Shale was being formed, some 530 million years ago.

Thus, for the really ancient fossils, we must look in the rock layers that were formed when the earth was relatively young. As can be done with amber inclusions, scientists studying the Burgess Shale fossils can piece together the ancient shallow seascape that existed when all life was confined to the ocean. Burrowing worms lived in the bottom sediment, trilobites and other creatures scoured the sea floor, and larger predators patrolled and fed on the rest. A general rule is that the farther you go back in time, the more distantly related the fossils are to modern animals. Thus, of all the creatures from the Burgess Shale, only slightly more than half show any relationship (and not a close one) with today's living animals. The remainder, such as *Hallucigenia,* with its seven pairs of pointed projections, have completely mystified scientists.

►

Desmond Collins (right) and George Poinar at the Burgess Shale campsite in Yoho National Park, British Columbia.

Through the kind assistance of Eric Langshaw and Rosemary Power, well-known naturalists who reside in Yoho National Park, I visited the Burgess Shale site during the summer of 1993. A helicopter that was bringing out a Japanese television crew agreed to carry me in and out of the area in order to meet Desmond Collins, the paleontologist in charge of the excavation of fossils. It was a wet summer, and low-lying clouds obscured the site most of the time. Thus it was touch-and-go when and if the helicopter pilot would risk the flight. Finally, a fresh wind partially cleared the clouds away and off we went. Flying over the mountains in one of those small bubble-type helicopters was an exhilarating experience—the closest I have ever felt to being a bird, gliding this way and that over the forests. I was to be dropped off at a rocky knoll about 1,000 feet across from the quarry. As the helicopter hovered over the spot, I repeated my instructions, and when the pilot waved his hand, I unbuckled my seatbelt, opened the door, grabbed my backpack, stepped on the footbar of the helicopter, jumped to the ground, reached up and closed the door, and then fell to my knees and waited for the helicopter to depart. The wind from the blades almost blew me off the knoll—there was nothing but pieces of shale to hold onto. After the helicopter left, I started to traverse the shale toward the quarry, where I could just barely make out three figures. The slope was so steep that I had to touch the ground almost continuously to keep from sliding down. It was slow going, and arduous. About halfway down it started to rain, and the shale became as slippery as a newly waxed floor. In that half hour it took me to cross to the site, I developed a deep respect for the mountain sheep and goats who were so well adapted to this environment.

At last, the quarry. I staggered up to the three workers and panted out my name. We chatted a bit after I caught my breath, and then they showed me where they were now digging and some of the fossils they had recently discovered. I had promised to meet Desmond Collins within two hours at his base camp some 2,000 feet down the slope, and not knowing how long that would take me, I decided not to dally. Thanking the three, I started sliding down the slippery pieces of shale toward the camp. Again I was leaning over so far that at one point I thought I might as well slide down the slope—but the jagged edges of the shale slabs discouraged me from that maneuver. By the time I could stand perpendicular to the ground, my thighs ached so much that I thought my legs would collapse. But I remained upright as I chatted with Des Collins and members of his team. We laughingly spoke about the possibility of recovering DNA from some of these strange animals and then about the reconstruction of a possible "Cambrian Park," with all of these creatures carrying on in their natural ways. There might be a remote chance for the former, but the reconstruction part was simply a wishful dream.

As I left this paleontologist's mecca by helicopter, I felt rejuvenated and inspired in a way that I imagine worshippers feel after an especially significant religious event. As we returned to San Francisco, our thoughts were focused on the *Jurassic Park* scenario. Now that biting flies were available in Cretaceous amber, would they contain some dinosaur DNA if they had fed on these creatures?

20   Older and Older

The publicity generated by our work with amber fossils continued to attract the attention of a variety of people. Some responded with news of their own amber projects. Out of the blue, one day in November 1991, a letter arrived from Ulf Christian Bauer of Schliersee, Germany, asking if I would like to receive some samples of old amber from the Bavarian Alps, amber that had been named Schlierseerit, after its locality.

When Bauer said old, he meant really old—some 225 million years old. After receiving my enthusiastic reply, he sent a small package containing tiny particles of dark, fragmented amber. It was the oldest amber I had ever seen, and its age placed it back in the Triassic period, around the time when the first dinosaurs appeared.

A cursory examination revealed no obvious insect or plant fossils, but there were small specks and fibers in one piece. A request to send more pieces with particles resulted in another shipment, fewer than before because, as Bauer emphasized, such pieces were very rare.

Over the next few months, Ben Waggoner—a paleontology graduate student at the University of California, Berkeley—and I examined those specks and fibers in detail. We discovered that they had structure and form and actually could be assigned to present-day modern groups of bacteria, algae, fungi, and protozoa. Spores were even embedded in the resin, and one sporelike organism had already germinated, possibly as a result of sugars present in the original resin.

This was the first time such life forms had been identified in Triassic amber. Bauer had actually been collecting amber from Raibler sandstone in the Leitnernase Mountains for eight years and had slowly amassed a collection of more than 2,000 pieces, ranging from one to eighteen millimeters in diameter. The site is located some 2,800 feet high in the beautiful foothills of the Bavarian Alps, with the amber occurring in an area only about six by eleven feet, along a hiking path. The amber is embedded in the sandstone, mostly as oval- or teardrop-shaped pieces, and the color varies from yellow to brown. The multiple shades of red probably reflect the different degrees of exposure the amber had to oxygen. But what kind of plant would have produced such ancient resin? We sent small clear samples to researchers who analyzed amber in order to determine their botanical source and place of origin. The methods of infrared spectroscopy and nuclear magnetic resonance were used. In both cases, the analyses produced spectra that were then compared with those of known samples of amber. Considering the age of the sample, it did not surprise us that the spectra of Bauer's amber differed from those of everything else.

Whatever the parent plant, it apparently did not survive as a resin producer. Some clues to the resin-producing plant were provided by leaf fossils in the sandstone that contained the amber fossils—one set of leaves belonged to a plant group known as the cycadophytes. Cycadophytes were strange plants but are considered by some people to be ancestors of present-day gymnosperms. They had leaves that resembled those of some fern plants today. Other fossil leaves encountered resembled those of an ancient ginko (*Baiera*), horsetails (*Sphenophyllum* and *Annularia*), and a conifer with leaves resembling those of *Araucaria (Walachea)*. The leaves of the resin-bearing plant were probably large, and had strong petiole bases. The petioles could have been grooved, thus forming little troughs at the bases. These troughs probably collected rainwater and remained filled for days and possibly weeks. Little by little, airborne spores settled into these stagnant micropools, some directly from the atmosphere and others washed down from the leaf surface or bark of the tree. Eventually, a small microcosm of organisms gathered in that water source. Green algae and certain protozoa could draw from the sun's energy. They and their metabolites released in the water furnished nutrients for sheathed bacteria and fungi. These in turn served as food for protozoa. Among the latter, and at the top of the food chain in this microcosm, were giant amoeba that glided through the water, engulfing their prey, digesting the soft parts and ejecting the harder portions.

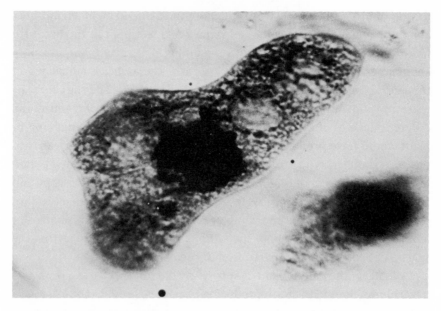

The oldest known naked terrestrial amoeba recovered from 225-million-year-old Bavarian amber.
(Specimen courtesy of Ulf Christian Bauer)

All of these events were occurring in a dense forest filled with various types of primitive plants but probably not yet containing any dinosaurs, only reptiles and amphibians. Suddenly, a flow of resin gushed out of the tree trunk, possibly from a wound. The avalanche covered everything instantaneously, halting all activities and preserving portions of that scene for some 225 million years. Branches of green algae were still in mats intertwined with long strands of sheathed bacteria. A possible spore had already begun to germinate. One amoeba was in the process of dividing, and another was just ejecting the undigestible portion of its last diatom meal. We had before us a flashphoto of life in a drop of Triassic water, preserved in three-dimensional form.

Particularly intriguing to us were the well-preserved nuclei, especially in the protozoa. Could there be DNA remaining in these small organelles, and if so, could we ever hope to recover some?

Most of these microorganisms turned out to be the oldest soft-tissue fossils ever known from a terrestrial environment, and that was exciting enough, but the resin's ability to preserve such delicate creatures for such a long time was even more astonishing.

What were the magic constituents in this ancient resin that could maintain the integrity of single-celled organisms? While attempting to answer this question, we learned just how little we know about the specific components of plant resins, and we know even less about which ones are responsible for their preservative properties.

Clues can be obtained from previous findings, however. In the first half of the twentieth century, and even earlier, kauri gum provided a flourishing business in northern New Zealand, as we discussed earlier in this book. The availability of not only kauri copal, but also of resin obtained directly from kauri pine trees, provided a more than adequate supply of material for scientific studies. Now we have evidence that Cretaceous amber from several sources around the world was derived from ancient kauri pines. Thus present-day kauri pine resin, assuming that it doesn't differ significantly from that of its ancestors, would contain all the components necessary to preserve organisms for millions of years. The results of numerous studies by several authors showed that kauri trunk resin contains the sugars, glucose, galactose, and arabinose. Such sugars would serve to withdraw the moisture from the bodies of entrapped organisms.

Thus, by the process of osmosis, water would pass from the organism into the resin, thus dehydrating the tissues by what we call inert dehydration. Once dehydrated, the tissues and cells have the potential to remain intact indefinitely, providing other conditions are maintained. Could the tissues also be preserved by chemical fixatives in the resin? Fixation could be accomplished by the alcohols (fenchyl, cis-communol and trans-communol), ethers (glycol ether), and terpenes that occur in kauri resin. Aldehydes are used currently to preserve modern tissues, and one oxygenated derivative of terpene hydrocarbons (found in kauri resin) is an aldehyde. The terpenes, as well as the alcohols and ether, would serve to inhibit microbial activity and autolytic processes that could break down the tissues of these inclusions. In addition, sealing off the organism with resin would greatly limit contact with oxygen and moisture, both of which would cause further decay.

What you have in the end is similar to a block of plastic containing an organism that has been fixed and dehydrated. Electron microscopists know that once so embedded, a tissue can remain in the drawer for years and retain its original integrity.

21 Famous Fossils

Publicity makes a fossil famous, but rarity, age, unique-
ness, and degree of preservation make a fossil valuable. The amber world
has experienced all types, and usually famous and valuable go together.

The Baltic amber fly containing nuclei was certainly a famous fossil.
The specimen itself was not rare—such flies are commonly found in many
amber deposits. The fossil was not extremely old; the great majority of Baltic fossils are dated around 40 million years of age. But the degree of tissue
preservation was exceptional, and it was demonstrated with the transmission electron microscope for the first time. That is what made this fly so
famous.

To place a monetary value on such specimens is difficult, because the
scientific value may or may not be correlated with price. However, at the
time of the discovery, *Discover* magazine published a story on the results,
and they wanted a well-known photographer on the East Coast to photograph the fly. They arranged to have someone hand-carry the piece from
Berkeley to New York and deliver it personally to the photographer. The
piece was insured for $25,000. Ironically, due to technical difficulties, the
photographs were not suitable for publication.

A second photographer, this one a sports photographer from Los Angeles, was flown up to Berkeley, bringing with him several trunks filled
with high-tech photographic equipment. The smallness of the amber piece
containing the fly surprised him, and none of his equipment proved suitable. In the end, *Discover* used a photo we took using a makeshift attach-
ment connecting an old Canon camera to an eyepiece of a student's
dissecting microscope.

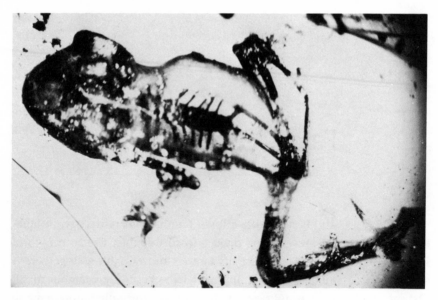

A rare frog in Dominican amber.

The *New York Times* nicknamed another famous amber fossil the "DNA bug." This was the weevil from Lebanese amber. Although rare and old, it became famous for its ability to yield the oldest known (120–135 million years) DNA.

Rare and scientifically important amber fossils sometimes turn up in unexpected places. A pair of fleas was discovered in a lady's amber pendant by an alert observer at a dinner party. Some twenty years ago, a couple from New Jersey showed a piece of Cretaceous amber they had collected to scientists at Harvard. Little did they realize that the piece contained two hymenopteran specimens that at the time were considered to be a link between nonsocial wasps and primitive ants.

Other famous amber fossils include a frog and a mushroom in Dominican amber, the best-preserved frog and mushroom fossils known. Although older remains of frogs do occur, the mushroom was probably the oldest of its kind. Both of these specimens appeal to collectors because of their rarity and degree of preservation. The condition of the frog told the story of how it most likely had ended up in a pool of resin. Its two broken legs showed signs of an aggressive attack on this poor creature, probably by a predatory bird. It was evidently caught by one leg, which subsequently broke as it struggled to free itself. The predator then grabbed it by another leg, probably shaking it back and forth in an effort to swallow it. At the last moment, the second leg broke and the victim slipped out of the predator's mouth and fell into a pool of resin. The bird then lost interest and flew off to catch something else. The bones of other frogs in the same piece of amber showed that this frog was not the first of its kind to be caught. Fly maggots surrounding the decaying bones indicated that scavengers frequented such nest areas.

In the case of the mushroom, delicate scales, gills, and even spores were still preserved in place. Obviously, a resin flow came down on this fragile organism like a tidal wave, engulfing all life in its path, including creatures that were inhabiting the mushroom as a home or temporary shelter. Beneath the cap was a colony of small invertebrates known as rotifers that were probably living within the gills. In fact, a few individuals were still wedged between the gills. Mites were living on the cap of the mushroom. As the resin covered the fungus, most of the mites jumped to safety. One didn't jump fast enough, and it was preserved in the amber, in the act of leaping from the rim of the cap.

Fossils like the one just described, which tell stories of past relationships (paleosymbiosis), are rare. They may represent common organisms, but they are involved in some commensal act and are caught in the middle of it. Scientists refer to this phenomenon as frozen behavior. Again this takes us back to the scenes of humans at Pompeii buried in the ashes of Mount Vesuvius. Many died in the act of performing a daily chore; some were in hospitals; others were aware of the danger and were fleeing for their lives. Can we relate this human tragedy to the same experience felt by insects trapped in amber?

In amber are preserved the daily life of invertebrates. There are fruit flies with phoretic mites still clinging to their bodies. These mites climbed on the flies, expecting to be carried to an exciting new habitat containing food and breeding sites—and instead they ended up immersed in resin. Entrapment in amber is a rare phenomenon. Even with ants and flies, which make up the majority of insects in amber, it is doubtful that any

A queen ant carrying a scale insect is an example of fossil symbiosis (paleosymbiosis) in amber.

selection pressure resulted in a natural avoidance of resin or in adaptations helping insects to remove themselves from it. Immersion in resin occurred suddenly, and no animal could prepare against it. Granted, creatures that had detachable appendages as a survival trick when grabbed by predators could save themselves by leaving these appendages in the resin. But immersion in resin was usually a one-way experience, and survivors were few.

Ants are common in amber, but ants in the process of carrying their young or a morsel of food back to the colony are rare, as are ants caught in the act of struggling to escape from a spider web. One could almost say that the essence of such creatures is preserved in the amber.

From other unique pieces we can learn what types of parasites had developed by certain time periods. Roundworm parasites fed on the blood and internal tissues of insects, now caught in the act of emerging from the bodies of their victims. There are pathogenic fungi that grew on insects and parasitic mites with their mouthparts still attached to their hosts' integument. What microscopic forms of parasites remain inside this vast array of invertebrates trapped in amber?

Perhaps, from the standpoint of fragility and size, one of the most spectacular finds we made was a colony of amoeba, caught in the middle of their cyclic events. Several were in motion with their pseudopods extended; another was in the process of dividing (mitosis); several others were voiding undigestible portions of diatoms from their bodies—all trapped in amber now 225 million years old!

In addition to famous and unique individual fossils, there are famous collections of fossils. One that I became personally involved with was the Brodzinsky-Penha collection of Dominican amber at the National Museum of Natural History, Smithsonian Institution, in Washington, D.C. A number of us wrote letters encouraging the Smithsonian to purchase this collection of 5,000-plus pieces, and everyone breathed a sigh of relief when it was finally acquired on May 9, 1985.

This collection represented a wide range of animal and plant fossils. Especially notable among the former was a bird feather and remains of a lizard and frog. The more unusual invertebrate fossils included a flea, portions of two dragonflies and a stonefly, a zorapteran (the first fossil zorapteran ever found), two walkingsticks, three male strepsipterans, two lacewings, centipedes, millipedes, pseudoscorpions, and a variety of rare beetles, wasps, and flies. Among the one-hundred-plus plant specimens were portions of mosses and liverworts, as well as seeds, petals, and leaves of angiosperms, including *Mimosa* leaflets and flowers.

At the time of purchase, the National Museum held only some 300 amber pieces in its fossil insect collection. After the acquisition, it could boast the largest collection of Dominican amber in North America, and probably in the entire world. The second largest at that time was in the Stuttgart National History Museum.

Asked to appraise the Smithsonian collection, I spent October 14–17, 1985, in the Natural History Museum examining the amber. My host was Don Davis, then chairman of the Entomology Department, who had been instrumental in acquiring the amber collection. Most of my time was spent with Gary Hevel, then in charge of the collection. The collection was still in the exact state in which it had been received—packed in forty-two old, rectangular shoe boxes. Each piece of amber was in a separate brown manila envelope that had its own identification number. All of the pieces had been polished, and they ranged from ten to sixty millimeters in length and from 0.5 to 10.0 grams in weight.

We estimated that, on average, each piece contained four separate plant or animal inclusions. Thus the 5,000 pieces actually represented some 20,000 inclusions. We decided that at least 20 percent of these specimens were holotypes (specimens that are designated as the type specimen for a new species). Holotypes are valued more highly than other individuals of the same species.

The examination was time consuming, but the Brodzinskys had done a very thorough job of amassing representatives of twenty-two of the twenty-six generally recognized insect orders, and it was a pleasure to see such beautiful specimens. The collection was appraised at $200,000 at the time, but the value has undoubtedly increased severalfold since then. In any case, a wonderful amber collection has been preserved for future generations of scientists.

Many rare fossils are purchased by private collectors for thousands of dollars, usually before anyone can organize funding for museum collections. We have tried to work with amber dealers in order to record rare finds before they became lost in private collections. Fortunately, many private collectors have loaned us rare specimens for official descriptions and documentation. This is important, because such pieces often are one of a kind and can provide information that may never be available again.

22   Amber Intrigue

Imagine: On the last day of your visit in Mexico City, a stranger walks up to you, reaches into his pocket, and flashes a large golden stone in your face.

"*Este ambar, señor,*" he says. "Look—my brother, he works in the mines in Chiapas, and every weekend he brings some pieces past the guards. That's why I can't take them to the stores. But this one has a lizard."

You stare at the large lizard in the center of the golden gem, and the mental wheels begin to turn. Since *Jurassic Park,* the price of amber has been going up, and vertebrates are known to be rare.

"How much?" you inquire.

"I only want seven hundred and fifty dollars, because my brother will lose his job if someone discovers he is taking the amber."

Your mind races: a vertebrate can sell for thousands of dollars—you could resell the piece in the States for ten times this price.

You purchase the piece, take it back to your hotel room, wrap it carefully in tissue, and bundle it between the shirts in your suitcase so it won't break. After a nerve-wracking trip, you arrive back home and carefully unpack the treasure, setting it under a layer of scarves in a lower drawer in the dresser. Days pass while you scan the newspaper, looking for someone who wants to buy amber fossils. Then, suddenly, a distressing article appears— on amber fakes and how they are becoming more common, especially in Mexico. You can distinguish many fakes, it reports, by using a simple salt-water test: dissolve two and a half level tablespoons of salt in a cup of water, for a salt solution that will float real amber. So, when no one else is around, you bring out your prize, make up the salt solution, and prepare

A dinosaur embedded in polyester surfboard resin to imitate amber.
(Prepared by Marlin Spike Werner and Sam Rubin, Pahoa, Hawaii)

to lower the amber into it. Is this necessary? It is probably authentic—it looks real—well, let's try anyway. The piece drops to the bottom of the cup like a lead weight, and your heart drops at about the same rate. You might say that you suspected it all along, at such a relatively cheap price. You could place the piece back in the drawer and try to forget the whole incident. However you react, rest assured that you weren't the first, nor will you be the last, to purchase a fake amber fossil.

Amber fakes are nothing new. I purchased my first fake in a shop in San Francisco in 1978. There were no fossils involved. It was a simple amber necklace composed of opaque yellow beads that looked like cloudy Baltic amber. I later discovered them to be 100 percent plastic.

In fake amber fossils, the insect itself is real but usually surrounded by plastic rather than amber. Less commonly, the organism again is real but placed inside a natural tree resin. The resin is harder to detect than plastic, because it has a piney smell—and it will float in salt solution. Its flaw is its softer composition, allowing it to be scratched and melted easily. In fact, it is the lower melting point that suits it for use in amber frauds: a hole can be melted out of the center, the organism inserted, melted resin poured around the specimen, and then the surface polished. A third type of fake can be more difficult to detect. A piece of real amber is cracked open, a small depression made in the center of the piece, an organism placed in the depression, and the two halves glued back together.

Plastic imitations of amber have existed since cellulose nitrate (celluloid) appeared in the late 1800s; and with the invention of lucite and epoxy resins, the forger now has a number of materials from which to choose.

Some of the most remarkable fakes were made with New Zealand kauri copal from the kauri pine, discussed earlier in this book. The practice of embedding insects in the material probably began as a way for gum diggers to alleviate boredom as they sat around the campfire, miles from civilization. After quitting the gum-digging trade, some veterans went on to perfect this practice, each developing his own special technique of melting the resin and inserting the lure. Sometimes it was a giant weta, sometimes a black wood cockroach; at other times, a large spider, a leaf, or even a green gecko or copper skink. Embedding lizards was in fashion for a time, and no one knows how many of New Zealand's native reptiles ended up embalmed in copal. The methods involved in the construction of these forgeries were kept secret and seem to have disappeared with their makers. Two or three such specimens come to our attention each year when sellers or purchasers want to know whether they are authentic amber.

Kauri copal, being a natural resin, has many of the same properties as amber—but not its hardness. There is usually a direct correlation between age and hardness, and kauri copal is relatively soft. Copal can be detected by melting a small area with a heated needle or polishing the piece with an electric sander. A resinous smell is obvious in both cases.

One of the best fake-amber samples I ever encountered originated in Germany. Small, irregularly shaped pieces of "amber" were made into necklaces and other types of jewelry, some containing insects. One day we received some of this material for verification. The matrix was fabricated with the sole purpose of passing tests for amber employed by the Gem Institute in Germany and the United States. Thus the imitation had a refractive index of 1.54, a specific gravity of 1.07, and an orange-yellow color. The pieces were somewhat shinier than real amber. Our most obvious clue came when we scraped the surface with a fine forceps: the color layer could be pulled away from the matrix, like the peel of an orange, leaving an opaque, dirty white material that bore no resemblance whatsoever to amber. Yet, because there were papers authenticating the material, the dealer initially refused to remove the merchandise from the market. Especially irritating to us was the practice of placing insects and mites within the matrix. If these pieces ended up in a museum collection, scientific conclusions regarding the appearance of select groups in the fossil record and future DNA studies would be completely false. Because of this, we initiated an investigation through the district attorney's office.

Our first task was to determine what this material was, if not amber. We attempted to dissolve the outer colored coat in all kinds of sophisticated solvents, but none of them touched it. Then fortune struck: A Ph.D. student who had borrowed a piece to identify a mite contained within, returned the next day, embarrassed. Apologetically she explained that, by accident, a drop of alcohol had fallen on the specimen, leaving a colorless opaque spot on the amber. She was shocked to see my glee over what to her was the partial destruction of a potentially valuable specimen. But we had overlooked the obvious. Ethanol was the only solvent we hadn't tested, and it told us that the outer layer was nothing other than shellac.

The importation of this material into North America from Germany was halted, but the manufacturer was not stopped. We eventually learned his address and, on one of our trips to Europe, passed through the town and even stopped to examine the establishment. From the outside, we could see a virtual chemistry production laboratory in the basement. We didn't bother to pay our respects.

Another amber-fraud case involved a large quantity of unpolished material from South America that had been sold as amber for jewelry and fossils. When the purchaser polished the material, however, it began to melt from the friction of the sander, and thus its use for jewelry was very limited. It had been sold as Baltic amber, but tests showed that it wasn't amber at all; it was a younger copal derived from modern legume trees from certain regions in South America. The material did contain very nicely preserved insects, and it could be useful in its own right to compare recent distributional changes in insect biodiversity in tropical rain forests. The matter was finally resolved out of court.

One unfortunate consequence of the amber hype is the increase in demand for this fossilized resin, accompanied by an increase in price (making fossils more difficult to obtain). Whenever the value of an item is increased, forgers, seizing the opportunity, come out of the walls and pedal their wares wherever they can. And if an insect in a piece of imitation amber increases the price ten to fifty times, imitations *will* appear.

Another line of intrigue involves attempting to verify localities of mysterious sources of amber and copal. At two such sites, in Kenya and in Colombia, the material tends to be lighter and softer than normal amber, and therefore suspiciously like copal. Although an age estimate can be determined through direct chemical analysis, obtaining index fossils from the strata in which the resin is embedded is also desirable, to determine at least a minimum age for the fossilized resin.

The Kenyan site reportedly is south of Mombasa, but all attempts of scientists to reach it have been thwarted. The last such effort was made by a professional paleontologist who apparently had reached Mombasa and was preparing to visit the mines the next day. As the story goes, he finished his supper at the dining tent and headed back toward his quarters. The path he took followed a riverbank, and at one point, on uneven ground, his foot slipped. Instantly, a crocodile seized it and pulled him down into the water. His cries brought help that saved his life, but he was in no condition to hunt for amber the next day. After a period of initial recovery, he flew back to the States and, as I heard it, never ventured in that part of Africa again.

In Colombia, humans pose the greatest dangers. Material is extracted from more than one site, and each site has its own characteristics. Reports we have received from various amber dealers are themselves secondhand, since none of the dealers has ever visited the sites. One site near Muzo (which happens to boast one of the largest emerald mines in the world) is supposed to be highly guarded and dangerous. The other locality is along the Magdalena River, deep in the rain forest. This, as well as sites to the north and east, is partly located in the cocaine growing and processing areas and apparently is frequented by Colombian guerrillas. Needless to say, only an invisible person would want to search for resin there. Within the past five years, two adventure seekers passing through Colombia volunteered to try to visit the sites and obtain information about them. We never heard from either of them.

Sometimes people accidentally discover an object that they believe to be amber but that later turns out to be something else. One recent case involved the 1988 discovery by a Swedish fishing boat of a huge lump (weighing over 200 kilograms) of "Baltic amber" from the bottom of the Baltic Sea. Found at a depth of seventy meters just east of the island of Gotland, the piece was purchased by Mr. L. Brost and exhibited in his private amber museum in Höllviken, Sweden. After the lump had sat indoors for some weeks, Mr. Brost noted that it began to shift its shape and "settle." Suspicious, he sent material for analysis to laboratories in the United States and in Poland—which identified it as the remains of a barrel of modern, commercial rosin derived from pines. Had it been authentic Baltic amber, it would have been the largest find ever reported.

◄

Leif Brost stands beside what he hoped was the largest known piece of Baltic amber. It turned out to be a barrel of modern pine resin that had been thrown into the Baltic Sea.

23   Ancient DNA

Amber fossils are not the only source of old DNA. In fact, they represent only one of many types of old samples that can yield nucleic acids. The term *ancient DNA* refers to DNA obtained from the remains of dead organisms. This includes a wide variety of topics over a wide expanse of time—from forensic examination of a murder victim whose body was found a week ago, to the fossil remains of an insect that died 125 million years ago. The term also refers to DNA obtained from the remains of humans during the historical period (the past 30,000 years), when events were recorded in writing or figures, as well as during the prehistoric period (back some 40,000 years), when the presence of *Homo sapiens* was determined by artifacts found in graves, middens, or living sites. In addition, *ancient DNA* includes DNA obtained from all of the fossils found in the earth's crust, including those in amber.

Scientists from around the world gathered at the second International Conference on Ancient DNA held at the Smithsonian Institution in Washington, D.C., in October 1993. During this three-day conference, speakers elaborated on methods of extracting ancient DNA, the significance and use of ancient DNA, and problems associated with the determination of ancient DNA.

What can be done with ancient DNA? First, if only a small portion of an extinct plant or animal is available, it may be possible to identify the organism from DNA extracted from the sample. Thus ancient DNA can be used to identify extinct species.

In addition to identifying fossil organisms when other characters are absent, ancient DNA can be used to arrange organisms in order of their

Russ Higuchi (left) and Raul Cano at the 1993 Ancient DNA conference in Washington, D.C.

Edward Golenberg (left), Hendrik Poinar (center), and Franco Rollo (right), all studying ancient plant DNA, at the 1993 Ancient DNA Conference in Washington, D.C.

first appearance in the past (known as phylogeny, or phylogenetic relationships). Ancient DNA can establish what types of sequences were present at a particular time in the past. It can also tell us how extinct organisms or races were related to present-day descendants. When dinosaur DNA is finally discovered, we hope it will tell us just how close some dinosaurs were to today's reptiles and birds.

Biogeographical questions can also be answered by ancient DNA analysis. Did Easter Island inhabitants originate from the western coast of South America, or were they part of a migration from southeastern Asia stepping across the islands of Oceania? Did the domesticated European rabbit arise from wild stock that spread out from northeastern Asia, or was it introduced by human activity from the western Mediterranean? Is the tree that produced the Dominican amber in the Caribbean more closely related to descendants in South America or to a relict species in East Africa?

These wide-ranging questions make the field of ancient DNA attractive to scientists in diverse fields. An increase in interest was shown by almost a tripling of participants at the second International Conference on Ancient DNA. Subscriptions to the *Ancient DNA Newsletter,* a means of communication for the members between meetings, have swelled to 600—not immense, but not bad for a field still in its infancy.

A number of unknowns remain in the field of ancient DNA. These questions nagged at early workers and are still matters of contention: What are the factors necessary for the preservation of DNA after an organism ends its life? How long can DNA remain in the bones of an animal that drowned in a river, sank in quicksand, was a victim of predation, or simply walked too close to the edge of a bubbling pool of tar? What happens to the DNA in chloroplasts when leaves fall naturally, or are stripped from the trees by galelike winds and fall into rushing streams or stagnant pools, or onto dry land, where they curl and break in fragments?

Organisms that fall into sticky resin and became immediately immersed are sealed away under very special and unique conditions. Thus far these circumstances are unrivaled for their successful preservation of organisms, and the factors responsible remain a mystery. The conditions appear to be more or less uniform, and preservation occurs rapidly. Thus there is little variation in relation to moisture, oxygen, and cellular breakdown, all of which affect the rate of DNA decay and are extremely unpredictable in other types of preservation.

A number of problems confront the ancient DNA researcher. Variable results from working with DNA remains, and determining whether the DNA recovered really is ancient and not just a modern contaminant, are two major difficulties. In science, an experiment that works once is interesting; an experiment that works twice could be believable; and one that works three times is probably valid. Reproducibility in experiments is important in any scientific field, especially in a newly founded one. But analysis of ancient remains for DNA presents special reproducibility problems. That only one in ten or one in fifty bones in a burial ground has retrievable DNA is quite possible, due to difference in the microenvironment of the burial habitat over hundreds or thousands or millions of years. Or, in a more extreme case, one in one thousand dinosaur bones from a particular site might yield DNA. Even if lucky enough to choose a positive sample, however, a scientist would have little chance of reproducing the results by using another bone from the same site.

The extreme sensitivity of the amplification reaction (the polymerase chain reaction) presents another major problem in dealing with ancient DNA. Unless directed otherwise, it will duplicate any DNA that enters the system, including contaminating DNA associated with the sample, DNA from airborne spores, or DNA from human skin cells. With such a serious risk of contamination, DNA researchers have developed methods of ensuring that only particular segments of the gene are amplified—segments that are characteristic and, if possible, unique to the specimen being studied. Such methods use specific primers that are based on an analysis of living relatives of the ancient species. Contamination is common in the handling of ancient human bones, because their sequences may closely resemble those of laboratory workers.

Some ancient samples are reluctant to yield what little remaining DNA they have. This is because the DNA is bound together with other compounds, known as inhibitors. In bones, the inhibitor may be a weak acid; in plant tissue, it may be a phenolic compound. These inhibitors usually must be removed before significant amounts of DNA can be retrieved.

Many researchers have remarked on how differently plant DNA behaves from animal DNA. Aside from certain regenerative properties of plant cells, some of these unique characteristics may mean that plant DNA is more resilient to decay than animal DNA. Plant cells, surrounded with a layer of cellulose, are more compartmentalized. They also have certain phenolic compounds that might bind to, and at the same time protect, large segments of DNA.

How can researchers working with ancient DNA repair damage that has occurred over time? The majority of ancient DNA can be assumed to have experienced some type of damage. We are told that DNA is constantly undergoing repair in the cells of our bodies, and that cancers and other degenerative diseases result from a breakdown in this repair system. The repair system begins to slow down as we age, so many aging effects can apparently be attributed to this phenomenon.

When an organism dies, the DNA no longer can employ a repair system, so the DNA molecule decays at rates dictated by the surrounding environment. What portions of ancient DNA can be repaired by inserting it into a living cell that has an active repair system in operation? Since repair enzymes are highly conserved, they are probably found in all living cells. If the DNA is totally destroyed, nothing can be done. However, if only partially destroyed, it might be repaired to some extent. This exciting prospect is a future challenge to molecular biologists.

The ancient DNA conference concluded with guarded optimism. Museums are full of ancient remains that might yield DNA—but little ancient DNA has yet been extracted. Whether this is because we don't know how to extract it, or because the great majority has already been destroyed, is difficult to say. The exception, of course, is fossils in amber: so far many well-preserved animal and plant fossils in amber appear to contain DNA.

That ancient DNA exists at all is astonishing. There is no survival value or evolutionary significance associated with the persistence of DNA after an organism dies. The survival of this molecule in substances like amber could well be considered a quirk of nature.

24   Future Implications

What have we discovered in amber? Individuals: spiders, leaves, beetles, and flies, and many others, each a unique scientific curiosity. But when amber specimens from one region are considered together, they become pieces of an ecological jigsaw puzzle; and when enough pieces are put into place, we can reconstruct the original habitat of a long-gone forest. Even more exciting is that the life forms in some amber deposits represent the remains of extinct tropical rain forests. When you consider how few records we have of past rain forests—because the climate and soil type of these ecosystems offer little opportunity for fossilization— the value of amber fossils becomes apparent.

Today's tropical rain forests cover only 7 percent of the earth's land surface, yet they contain more than half of the animal and plant species in the world. At the current rate of land reform, these forests will disappear within the next century, resulting in the extinction of hundreds of thousands of species. To imagine the impact of such a scenario on the human race can be difficult. Some people wonder why we should worry about rain forests at all when we know so little about their plant and insect species, many of which haven't even been named. That fact in and of itself is one reason that great numbers of people (especially scientists) work to save the forests—to study the organisms and to learn something of how they survive and interact. And then there are the larger questions: What climate shifts will occur when the rain forests are eliminated, and how will these affect the remaining life on the planet? Will the new balance be to humans' detriment? Is this just one step in the process of the earth's transformation into a new world that will host only the most hardy life forms?

Hand in hand with biodiversity goes genetic diversity, or genetic variation. By this we mean the range and amount of genetic makeup in a species. A normal species comprises several populations that breed mostly among themselves but occasionally interbreed with members of neighboring populations. Each population has its own genetic makeup, which differs slightly from that of neighboring populations. If neighboring populations are eliminated, the original population can no longer receive new genetic material from outside; thus, genetic diversity is reduced.

A broad range of genetic diversity—essential to the long-term survival of life—is restricted when a species shrinks to a small population. Referred to as a "bottleneck" by population biologists, this reduction means that a species may not have access to the genetic diversity that would ensure its survival—it may not survive a climatic change, a new predator, or a modification of its food source.

An example involving the search for high-yielding crops for Third World countries illustrates this point. In the Green Revolution program some years back, plant geneticists selected a small number of crops that would produce high yields. In so doing, they reduced the genetic variation within the plant species to such a degree that little natural resistance to pests and diseases remained, and the plants depended largely on fertilizers. As long as a minimal enrichment and pesticide program was in effect, the yields were high and the farmers cast aside their old seeds, believing they would never need them again. But it became expensive and time consuming to fertilize and spray these crops. When the process ended, their reduced adaptive, self-sustaining ability became shockingly apparent. The population crashed, and the farmers searched frantically for the old seeds they had cast aside.

How does the importance of genetic diversity in a species relate to our concern with rain forests and amber? Many plants being destroyed in today's rain forests are wild relatives of domestic crops, as emphasized by Paul and Anne Ehrlich in their book *Earth,* and we might learn to incorporate ancient genetic sequences from plant remains in amber to increase the genetic diversity of crops. The remains of plants representing two major food groups in the world today—legumes and grasses—occur in amber; perhaps these fossils could provide sequences that would increase the yield, or control the production of defensive compounds, when introduced into crops.

Norman Farnsworth estimates that nearly 120 chemicals from higher plants are used by almost 4 billion people worldwide for medicinal purposes. There is a record of *Taxus* (yew plant) in Baltic amber, as well as a wide range of other plants, possibly some with medicinal properties. Could we isolate the gene sequences responsible for producing an ancient taxol-like substance? Lower life forms, such as bacteria, actinomycetes, and fungi, also occur in amber. We may be able to recover DNA sequences that would control the production of novel antibiotics from these microorganisms.

Amber contains a range of bloodsucking arthropods whose descendants today carry pathogens that infect humans and domestic animals. Mosquitoes transmit viruses, protozoa, and filarial worms. Also present are biting midges, sandflies, horseflies, fleas, blackflies, and ticks. Representatives of these groups are known to carry viruses, tularemia, anthrax Bacillus, protozoa, rickettsia, plague, murine typhus, and organisms causing encephalitis, hemorrhagic fever, scrub typhus, and Lyme disease. Early evolutionary stages of these parasites might be used to provide a basis for producing vaccines.

In amber we effectively have zoos and botanical gardens of past worlds, accompanied with the remains of the DNA of their components. Just as zoos and gardens today provide a source of genetic diversity of species, so does amber preserve a portion of the genetic information of extinct organisms.

What information can actually be obtained from amber inclusions? We know we can extract small fragments of DNA (gene sequences) from insects and plants in amber. And although the possibility of re-creating extinct creatures from such small fragments is nil, these small sections of DNA can be compared with corresponding segments in their modern descendants. Such comparisons can verify the identification of the extinct organism and tell us how close it is to present-day related species, thus helping to establish a line of descendance. If the rate of certain changes in the portion of the recovered genetic code is known, these ancient sequences can be used as a molecular clock to estimate the time necessary for the separation of species. This information can also help us to understand how species ended up, geographically, where they are today.

Spores are a remarkable biological phenomenon. Although not actively functioning, and metabolic activity in them is often not detectable, spores represent life because under the right conditions (usually temperature and moisture) they can reinitiate active development. This ability of spores, and even some stages of multicellular invertebrate animals, to enter a completely dormant period has intrigued scientists for ages. Lacking detailed observation, researchers offered the theory of spontaneous generation, or the idea that organisms in decaying matter arose spontaneously from nonliving elements in the medium. When spores were first discovered, they inspired the philosophical question of what is alive and what is dead.

To many people, differentiating between life and death appears straightforward. Normally, organisms that do not respond to stimuli, do not assimilate, or excrete, are dead. But spores have all of these characteristics and they are not dead—they are merely resting.

There are two types of death in humans: clinical death, involving an absence of heartbeat or breathing; and brain death, where the brain and brain stem irreversibly cease all functions. In brain-dead persons, some cells and tissues remain alive for a period. An organ can be removed and transplanted into another individual, thus continuing the survival of part of the original individual. If and when organs no longer function, a particular genetic line can be continued through cell culture.

The epitome of cell longevity is perhaps the famous HeLa cell line, initiated from cervical cancerous cells removed from a thirty-year-old woman in 1951. Leonard Hayflick and others have shown that many normal cells in the human body will divide for a time, and then either stop dividing or die. Much has been written about the factors responsible for cell death, but little is known about the mechanisms involved. The HeLa cells had been transformed and no longer are subject to the regimented code of normal cells; thus they continue to divide in tissue culture, and have been doing so for the past forty-plus years. Although some scientists claim that these cells now differ from human cells and represent a new type of organism, genetic analysis should show that they are modified human cells that seem to have become immortal, as long as people feed and nurse them. If these cells for some reason did revert back to normal cells, they would eventually stop dividing and die. Would that be the end of the line? It would, unless the DNA from those cells could be preserved in their own right. Would this still represent life?

Today we are faced with the question of whether there can be a molecular level of life. Because it is possible to insert isolated DNA sequences into cells and obtain physical characters (expression), these isolated DNA sequences could be considered alive at the molecular level. The definition of life therefore could be expanded to include any entity that can grow and/or reproduce when placed in an optimum environment, or any entity that can direct protein synthesis pathways expressed by itself or in another organism.

We have not yet established whether fossils from all amber sources hold retrievable DNA, but we suspect that they do. If DNA obtained from fossils in amber can be expressed in modern-day descendants, then life can indeed endure for eons.

Bibliographic Synopsis

1. Amber from the Sea

More on the basic aspects of early amber trade routes can be found in the work *The Ancient Amber Routes and the Geographical Discovery of the Eastern Baltic,* by Arnolds Spekke (1957, M. Goppers, Stockholm).

2. The Russian Connection

Details on searching for the amber room were included in a 1992 article in *People* magazine titled "A Treasure in Amber"; in John Ross's article on amber in the January 1993 issue of *Smithsonian*; and in the book *Quest: Searching for Germany's Nazi Past: A Young Man's Story,* by I. Melchior and F. Brandenburg (1990, Presidio Press, Novato, California).

3. Looking at Amber Up Close

Additional information on nematodes can be obtained from my 1983 book *The Natural History of Nematodes* (Prentice-Hall, Englewood Cliffs, New Jersey). The topic of dormancy and the structure and formation of spores can be found in most texts on microbes, such as *Microbiology*, by R. Cano and J. Colomi (1986, West Publishing Co., St. Paul, Minnesota).

4. African Safari

For a popular account of the disease causing river blindness, see my article "Rivers of Darkness," in *Pacific Discovery* (1979, vol. 32, no. 4). Further information on African army ants can be found in *The Ants,* by B. Hölldobler and E. Wilson (1990, Belknap Press, Cambridge, Massachusetts).

5. Adventures with Tomb Amber

The original article on the use of infrared spectra to determine the origin of Minoan and Mycenaean tomb amber was written by Curt Beck in 1966 (*Greek, Roman and Byzantine Studies,* vol. 7, pp. 191–211). An illustrated account of tomb amber artifacts in the British Museum was published by D. Strong in 1966 as a *Catalogue of the Carved Amber in the Department of Greek and Roman Antiquities* (The Trustees of the British Museum).

6. Excursion in Poland

An illustrated guide (in English) to the amber exhibition in the Museum of the Earth in Warsaw, *Amber in Nature,* was prepared by Barbara Kosmowska-Ceranowicz and her colleagues in 1984. Many early (fifteenth- to eighteenth-century) European amber carvings are illustrated in the *Catalogue of European Ambers in the Victoria and Albert Museum,* by Marjorie Trusted (1985, Victoria and Albert Museum, London).

8. Bringing Them Back Alive

Procedures used in our laboratory for working with unknown microorganisms were discussed in the appendix of our original book (with G. Thomas), *Diagnostic Manual for the Identification of Insect Pathogens* (1978, Plenum Press, New York), which was later expanded into the *Laboratory Guide to Insect Pathogens and Parasites* (1984, Plenum Press). How stingless bees occur in amber is discussed in my 1992 article "Fossil Evidence of Resin Utilization by Insects," in *Biotropica* (vol. 24, pp. 466–468), and in my *Life in Amber* (1992, Stanford University Press).

9. Million-Year-Old Cells

Our original article on the ultrastructure of tissue in the famous fly appeared in 1981 in the *IRCS Medical Sciences* (vol. 9, pp. 673). The more detailed work appeared in *Science* (vol. 215, pp. 1241–1242 in 1982. Additional photographs and commentary were presented in our 1985 article "Preservative Qualities of Recent and Fossil Resins: Electron Micrograph Studies on Tissue in Baltic Amber," which appeared in the *Journal for Baltic Studies* (vol. 16, pp. 222–230). A popular account of our work was written by J. Gorman for *Discover* magazine in 1982 (vol. 3, no. 5, pp. 36–38).

10. Pursuing Ancient DNA

The first article on obtaining DNA from the quagga was published by R. Higuchi and others in 1984 in *Nature* (vol. 312, pp. 282–284). In the following year, S. Pääbo published on the molecular cloning of DNA from an Egyptian mummy (*Nature,* vol. 314, pp. 644–645). The initial studies on attempts to extract DNA from amber insects appeared in 1984 in *Federation Proceedings* (vol. 43, pp. 1557). The idea of the polymerase chain reaction was written up by its discoverer, Kary Mullis, in the April 1990 issue of *Scientific American.*

11. Amber South of the Border
A map of the Chiapas amber mines was presented in my book *Life in Amber* (1992, Stanford University Press). Historical observations on pre-Columbian amber trade were made by F. Blom in 1959 in a publication of the Hamburg Museum of Folk Art entitled "Historical notes relating to the pre-Columbian amber trade from Chiapas" (vol. 29, pp. 24–27). An account of the Lacondonian Indians and their plight appears in the book *Gertrude Blom—Bearing Witness,* edited by A. Harris and M. Sartor (1984, University of North Carolina Press, Chapel Hill).

13. Western European Collections
C. Baroni Urbani published his papers on the first fossil gardening ants, the first fossil *Anochetus* ant, and the first fossil *Leptomyrmex* in the *Stuttgart Studies on Natural Science* in 1980. The first fossil aphid from neotropical amber was published in the journal *Psyche* (vol. 95, pp. 153–165), by O. Heie and myself, in 1988. An excellent book on Baltic amber fossils was published by S. Larsson in 1978 (Scandinavian Science Press Ltd., Klampenborg, Denmark). D. Schlee has issued several booklets on various aspects of amber and amber fossils, all published in Germany by the Stuttgart Natural History Museum.

14. Amber in the Caribbean
The account of the movement of continental plates containing resin-producing trees was taken from my article "The Amber Ark," which appeared in *Natural History* in 1988 (vol. 97, pp. 42–47). The arrival of Columbus in Hispaniola and the history of the Dominican Republic are described by the late Carlton Rood in his book *A Dominican Chronicle* (1985, Editora Corripio, Santo Domingo, Dominican Republic). The various amber mines in the Dominican Republic are discussed in my book *Life in Amber* (1992, Stanford University Press). My description of the Dominican Republic amber tree appeared in the journal *Experientia* in 1991 (vol. 47, pp. 1075–1082).

15. Amber Down Under
The New Zealand bats, as well as other introduced mammals, are discussed in the book *Collins Guide to the Mammals of New Zealand* (1986, Collins, Auckland), by M. Daniel and A. Baker. Features of

the tuatara and liopelmid frogs are presented in Joan Robb's *New Zealand Amphibians and Reptiles* (1980, Collins, Auckland). Characteristics and ranges of the kiwis are discussed in the *Collins Guide to the Birds of New Zealand* (1981, Collins, Auckland), by R. Falla, R. Sibson, and E. Turbott. Wetas, along with other endemic and exotic New Zealand insects, are described in D. Miller's book *Common Insects in New Zealand* (1984, revised by A. K. Walker, Reed Ltd., Wellington). The story of the New Zealand Kauri pine is presented by J. Halkett and E. Gale in their book *The World of the Kauri* (1986, Reed Methuen Ltd., Auckland). A colorful history of digging for kauri copal can be found in A. Reed's book *The Gumdiggers, the Story of Kauri Gum* (1972, Reed Ltd., Wellington). Mary White's description of *Agathis jurassica* appeared in the *Records of the Australian Museum* (1981, vol. 33, pp. 695–721).

16. Rekindling the Quest for Ancient DNA

The first paper dealing with the isolation of DNA from an amber insect, "Isolation and Partial Characterization of DNA from the Bee *Proplebeia dominicana* (Apidae: Hymenoptera) in 25–40-Million-Year-Old Amber, was authored by R. Cano, H. Poinar, and me. This was followed by a second paper, "Enzymatic Amplification and Nucleotide Sequencing of Portions of the 18s rRNA Gene of the Bee *Proplebeia dominicana* (Apidaer: Hymenoptera) Isolated from 25–40-Million-Year-Old Dominican Amber," by R. Cano, H. Poinar, D. Roubik, and me. Both articles appeared in *Medical Science Research,* in the April 1, 1992, issue and in the September 1, 1992, issue respectively. On September 25, 1992, the paper "DNA Sequences from a Fossil Termite in Oligo-Miocene Amber and Phylogenetic Implications," by R. DeSalle, J. Gatesy, W. Wheeler, and D. Grimaldi, appeared in *Science.* DNA from the extinct Dominican Republic amber plant appeared in *Nature* (1993, vol. 363, pp. 677) as "Oldest DNA from Plants," by H. Poinar, G. Poinar, and R. Cano.

17. Going Back in Time

The principle of uniformitarianism has been around for a while but was thoroughly documented for the fossil record by A. Boucot in his book *Evolutionary Paleobiology of Behavior and Coevolution* (1990, Elsevier Science Publishers, Amsterdam). The Lebanese amber collection of A. Acra was commented on by P. Petit-Roulet in *The New Yorker* magazine (June 28, 1993, p. 31). The famous Lebanese amber weevil was described by Kuschel and me as *Libanorhinus succinus* in *Entomologica Scandanavica* (1993, vol. 24, pp. 143–146). The paper presenting the amplification and sequencing of DNA from this 120–135-million-year-old weevil entitled

"Amplification and sequencing of DNA from a 120–135-million-year-old weevil," appeared in *Nature* (vol. 363, pp. 536–538), authored by R. Cano, H. Poinar, N. Pieniazek, A. Acra, and me.

18. *Jurassic Park* Repercussions

The science fiction story about bringing dinosaurs back to life from their DNA found in the bodies of bloodsucking insects preserved in amber was Michael Crichton's *Jurassic Park* (1990, Alfred A. Knopf, New York). Ted Pike's discovery of the first mosquito in Cretaceous amber was mentioned in my article "Insects in Amber," in the *Annual Review of Entomology* (1993, vol. 38, pp. 145–159). Mention of biting midges in Cretaceous Canadian-Alberta amber appeared in an article I wrote with T. Pike and G. Krantz, "Animal-Animal Parasitism" (*Nature*, 1993, vol. 361, pp. 307–308).

19. Amber from the Dinosaur Period

The types of dinosaurs found in Alberta, Canada, are discussed in Ron Stewart's *Dinosaurs of the West* (1988, Lone Pine Pub., Edmonton) and Monty Reed's *The Last Great Dinosaurs*. The first and most complete published study done on Canadian amber was authored by F. Carpenter and colleagues, "Insects and Arachnids From Canadian Amber" and appeared in the *Geological Series of the University of Toronto Studies* (1937, vol. 40, pp. 7–62). Additional information on Canadian amber was presented by J. McAlpine and J. Martin in the article "Canadian Amber—A Paleontological Treasure Chest," published in the *Canadian Entomologist* (1969, vol. 101, pp. 819–838). A concise and informative account of the Burgess Shale fossils by H. Whittington appeared in the book *The Burgess Shale* (1985, Yale University Press, New Haven).

20. Older and Older

The discovery of terrestrial soft-bodied microorganisms in Triassic amber was presented by me, B. Waggoner, and U.-C. Bauer in *Science* (1993, vol. 259, pp. 202–204). We also described the earliest known soft-bodied amoeba from these deposits (1993, *Naturwissenschhaften*, vol. 80, pp. 566–568).

21. Famous Fossils

Famous and rare fossils are figured in my book *Life in Amber* (1992, Stanford University Press). Also in the above is a discussion of the various types of symbiotic relationships found in amber. I and D. Canatella published on the famous frog in *Science* (1987, vol. 237, pp. 1215–1216). The only mushroom in amber (*Coprinites dominicana*) was published in *Science* by me and R. Singer (1990, vol. 248, pp. 1099–1101).

22. Amber Intrigue

An article I wrote, in *Gems and Minerals* (1982, no. 534, pp. 80–84) describes how to distinguish real from fake amber. It also includes photographs of different types of fake material, including the material from Germany. The analysis on the large mass of resin removed from the Baltic Sea, written by C. Beck, E. Stout, and B. Kosmowska-Ceranowicz, entitled "A large find of supposed amber from the Baltic Sea," appeared in the Swedish journal *Geologiska Föreningens i Förhandlingar* (1993, vol. 115, pp. 145–150).

23. Ancient DNA

The first book on ancient DNA, entitled *Ancient DNA,* appeared early in 1994. It is edited by B. Herrmann and S. Hummel and published by Springer-Verlag, New York. Its chapter on DNA from amber inclusions was written by me, H. Poinar, and R. Cano. A more popular article on ancient DNA by S. Pääbo appeared in the November 1993 issue of *Scientific American.*

24. Future Implications

Information on biodiversity was obtained from the book by the same title edited by E. Wilson (1988, National Academy Press, Washington, D.C.). This work includes a chapter, by N. Farnsworth, dealing with plant medicines. Other views on this topic were obtained from Paul and Anne Ehrlich's book *Earth* (1987, Franklin Watts, New York). My views on life at the molecular level were presented earlier in the article "Still Life in Amber," which appeared in *The Sciences* (March/April 1993, pp. 57–68).

About the Authors

George Poinar is on the faculty of the Entomology Division at the University of California, Berkeley. He received his B.S., M.S., and Ph.D. degrees in the biological sciences from Cornell University. His background in zoology and botany proved invaluable for his later studies on amber inclusions. George Poinar's photographs of fossilized insects and plants in amber have appeared in scientific journals, newspapers, magazines, and textbooks around the world. With each subject, Poinar strives to record not only its physical characteristics, but also its artistic qualities. His publications include more than 300 technical papers, scientific books, popular articles, and short stories.

Roberta Poinar attended the University of California, Berkeley, where she took a part-time job as an electron microscopist. After receiving her Bachelor's degree, she remained at Berkeley, working as an electron microscopist in the Departments of Zoology, Nutrition, and Entomology. She is the author or coauthor of some ninety scientific publications, covering a wide range of topics involving electron microscopy.

The authors began working together in 1970 when Roberta became the electron microscopist for the Insect Pathology Unit in the Department of Entomology at U.C., Berkeley. Several years later, they began their first amber project, and they have worked together ever since.

Index

Photography Credits

Photographs not otherwise credited were taken by George and Roberta Poinar.

Cover photograph (winged termite) reprinted from *Life in Amber* by George Poinar, with the permission of the publisher Stanford University Press, © 1992 by the Board of Trustees of the Leland Stanford Junior University.

Color plates of ant bug and mushroom, and the photograph on page 21, reprinted from *Life in Amber* by George Poinar, with the permission of the publisher Stanford University Press, © 1992 by the Board of Trustees of the Leland Stanford Junior University.

Photographs on pages 127, 128, and 129 by the Northwood Brothers, courtesy of Otamatea Kauri and Pioneer Museum, Matakohe, New Zealand.

Photograph on page 61 by Glen Epling.

Photographs on pages 66 and 118 reprinted with permission by Pat Craig.

Photograph on page 71 courtesy of Leona G. Wilson.

Photographs on pages 80, 83, and left photo on page 88 reprinted with permission by J. Wyatt Durham.

Photographs on pages 60 and 93 reprinted with permission courtesy of *The Extinct DNA Newsletter,* Number 3, April 1983.

Photograph on page 93 by Karen Tkach.

Photograph on page 140 by Hendrik Poinar.

Photograph on page 160 reprinted with permission by Greg Poinar.

Painting on page 40 by Ludwik Leszko, of Cracow; courtesy of Barbara Kosmowska-Ceranowicz, Museum of the Earth, Warsaw, Poland.